Exploring the Language of Adventure Tourism

Language and Text Studies

Edited by
Alexander Brock and Daniela Pietrini

Volume 24

Berlin · Bruxelles · Chennai · Lausanne · New York · Oxford

Isabel Durán-Muñoz / Eva Lucía Jiménez-Navarro (eds.)

Exploring the Language of Adventure Tourism

A Corpus-Assisted Approach

Berlin - Bruxelles - Chennai - Lausanne - New York - Oxford

Library of Congress Cataloging-in-Publication
A record in the CIP catalog has been requested for this book of the Library of Congress.

Bibliographic Information published by the Deutsche Nationalbibliothek
The Deutsche Nationalbibliothek lists this publication in the Deutsche Nationalbibliografie; detailed bibliographic data is available online at http://dnb.d-nb.de.

This edited volume has been carried out within the framework of the R&D project "DicoAdventure: diseño y desarrollo de un recurso electrónico especializado bilingüe (inglés, español) sobre el turismo de aventura a partir de marcos semánticos" (Ref. UCO-1380857-F), co-funded by the Operational Programme FEDER 2014-2020 and the Consejería de Economía, Conocimiento, Empresas y Universidad of the Andalusian regional government.

ISSN 0941-4134
ISBN 978-3-631-88010-4 (Print)
E-ISBN 978-3-631-88677-9 (E-PDF)
E-ISBN 978-3-631-88678-6 (EPUB)
DOI 10.3726/b 21581

© 2024 Peter Lang Group AG, Lausanne
Published by Peter Lang GmbH, Berlin, Deutschland

info@peterlang.com - www.peterlang.com

All rights reserved.

All parts of this publication are protected by copyright. Any utilisation outside the strict limits of the copyright law, without the permission of the publisher, is forbidden and liable to prosecution. This applies in particular to reproductions, translations, microfilming, and storage and processing in electronic retrieval systems.

This publication has been peer reviewed.

Foreword

Once upon a time, *tourism* was a socio-economic phenomenon that developed massively after the Second World War in terms of two main aspects: leisure in holiday resorts and visits to monuments and cities. In accordance with these two trends, different lexical choices were incorporated into the language of tourism, coming from sectors such as geography, history, art history, gastronomy, and folklore, together with a hard core of terms relating to the management of travel agencies, transport, and the hotel industry. Likewise, specific textual genres were consolidated, such as the travel guide and the promotional brochure, responsible for describing and extolling the wonders of the tourist destination, guiding the traveller's gaze. This preponderance of the visual elements characterises early tourism promotion, as highlighted in one of the best-known studies on tourism and its motivations, *The tourist gaze*, by John Urry (1990). In the face of the centrality of art and landscape, the figure of the tourist remains in the background (Febas Borra, 1978).

Over time, the focus shifts from the place visited to the visitor: the idea of tourism as contemplation is replaced by tourism as practice, and *experience* becomes the keyword. John Urry himself lavishly illustrates this shift in perspective in a more recent work, written together with Jonas Larsen, *The tourist gaze 3.0* (Urry & Larsen, 2011). At the same time, tourism diversifies into different types, in which certain sectors of activity systematically converge; as a way of illustration, wine tourism develops thanks to the confluence of the wine segment and tourism (Luque Janodet, 2022). It goes without saying that the enormous growth of digital tourism, which boosts the autonomy of the tourist in the search for information and makes them the provider of tourist communication, favours the expansion of the different types of tourism, around which nomenclatures, lexical combinations, and specific discursive forms are built, as well as communities of travellers who share interests, experiences, and evaluations. In this way, tourism opens up into a wide range of *tourisms*, with their own profiles and motivations.

The production of specific lexical and discursive features varies greatly from one type of tourism to another, depending on the activities specific to each of them. In the case of *business tourism*, for example, tourist attractions act as a corollary of professional activity whose communicative patterns do not fall within the scope of tourism. For *adventure tourism*, on the other hand, the practice of certain sports, with special reference to those that take place outdoors, is fundamentally constitutive, as can be inferred from the definition of this type of tourism offered by the World Tourism Organisation (UNWTO) (2023):

> Adventure tourism is a type of tourism which usually takes place in destinations with specific geographic features and landscape and tends to be associated with a physical activity, cultural exchange, interaction and engagement with nature. This experience may involve some kind of real or perceived risk and may require significant physical and/or mental effort. Adventure tourism generally includes outdoor activities such as mountaineering, trekking, bungee jumping, rock climbing, rafting, canoeing, kayaking, canyoning, mountain biking, bush walking, scuba diving. Likewise, some indoor adventure tourism activities may also be practiced.[1]

The continuous flourishing of new sports and the consequent coining of terms to define them attest to the close link between tourism and sports activities that involve immersion in nature and the perception of the landscape through active participation. The UNWTO definition provides us with other relevant elements, such as the idea of risk and physical or mental effort. Therefore, a figure of tourists is presented here in a very different way compared to mass tourists, who let themselves be led like a child by the hands of agencies and guides. It should also be noted that the risk can be "real or perceived", which gives rise to discursive constructions in which, beyond the real risk involved in the activities carried out and the physical preparation they require, this perception is enhanced, as is clearly seen in the texts aimed at promoting this type of tourism. Every tourist can feel free to cross the threshold that turns a simple hike into an adventure, putting themselves to the test without taking any real risks. Similarly, the language of this tourism segment tends to nurture the perception of a departure from everyday to unusual, isolated, or remote places, and the emotions they arouse.

The works collected in this volume are based on textual corpora and aim to describe, through various theoretical approaches, the lexical and discursive specificities of adventure tourism, with a view also to providing useful resources for translation and terminology, between language pairs such as English-Spanish and English-Italian. A significant part of them uses the *DicoAdventure* project as starting point, which was coordinated by Isabel Durán-Muñoz and within whose framework a Spanish-English terminological resource was created. Additionally, the specialised bilingual (English-Spanish) corpus ADVENCOR, compiled in this context (cf. Durán-Muñoz & Jiménez-Navarro, 2021), is relied

---

1 https://www.unwto.org/glossary-tourism-terms#:~:text=adventure%20tourism%20generally%20includes%20outdoor,activities%20may%20also%20be%20practiced [Last accessed: 03/10/2023].

upon. The *DicoAdventure* dictionary combines the approaches of Explanatory Combinatorial Lexicology and Frame Semantics: a theoretical approach that allows the core elements related to the adventure activities (e.g., place where they are carried out, the instrument they need, the people involved, etc.) to be incorporated into the definitions, with tangible benefits for the characterisation of the specialised lexicon of adventure tourism. In addition to this body of works, there are others, also based on corpora, which explore different features of this speciality of tourism, such as the case of accessibility in this discourse. In short, the volume offers a unitary set of valuable contributions focusing on the linguistic forms used in the various texts, like verbs, mainly of motion, and specialised adjectives, as well as collocations and phraseological units expressing knowledge relevant to the type of activity being described; cognitive semantics provides a solid theoretical foundation for most of them.

I consider this approach to specialised terminology, open to phraseological units and grammatical categories other than nouns, such as adjectives and verbs, to be particularly appropriate in the case of the language of tourism, which places the features of this segment in a coherent and innovative framework. In essence, a dynamic vision of linguistic resources is presented, in line with the key role of experience in tourism today, in three languages of great importance for this sector, that is, English, Spanish, and Italian. From a cross-linguistic and intercultural point of view, both common elements and differences are observed, and it is not surprising that, compared to the English texts, in the Italian ones the point of arrival is more important than the route, as a burden of a conception of tourism centred more on contemplation than on the active role of the tourist.

From the perspective of my lifelong commitment to the language of tourism, which dates back to the time when this area of the language had hardly any dignity as a specialised language, I can only highlight the depth of the chapters collected here and the high level of professionalism which they are the result of. This is a hopeful sign for future developments in an area which, due to its links with a thriving industry of activity (despite the ensuing hiatus that was brought on by COVID-19), is characterised by an enormous textual production and a growing demand for translation. Among other future tasks, I would point out that of describing the use made of the specialised units by tourists in the texts they produce themselves. I am especially grateful to the editors of the volume, Isabel Durán-Muñoz and Eva Lucía Jiménez-Navarro, for offering me the pleasant task of providing the foreword.

Maria Vittoria Calvi

References

Durán-Muñoz, I., & Jiménez-Navarro, E. L. (2021). Colocaciones verbales en el turismo de aventura: Estudio contrastivo inglés-español. In G. Corpas Pastor, M.ª R. Bautista Zambrana, & C. M. Hidalgo-Ternero (Eds.), *Sistemas fraseológicos en contraste: Enfoques computacionales y de corpus* (pp. 121–142). Comares.

Febas Borra, J. L. (1978). Semiología del lenguaje turístico (Investigación sobre los folletos españoles de turismo). *Estudios Turísticos, 57-58*, 17–203. https://doi.org/10.61520/et.57-581978.418

Luque Janodet, F. (2022). Hacia una caracterización del discurso de la cata de vino como lengua de especialidad. *Onomázein, 58*, 1–17. https://doi.org/10.7764/onomazein.58.01

Urry, J. (1990). *The tourist gaze: Leisure and travel in contemporary societies*. Sage.

Urry, J., & Larsen, J. (2011). *The tourist gaze 3.0*. Sage.

Acknowledgements

This edited volume has been carried out within the framework of the R&D project "DicoAdventure: diseño y desarrollo de un recurso electrónico especializado bilingüe (inglés, español) sobre el turismo de aventura a partir de marcos semánticos" (Ref. UCO-1380857-F), co-funded by the Operational Programme FEDER 2014–2020 and the Consejería de Economía, Conocimiento, Empresas y Universidad of the Andalusian regional government.

We would also like to thank the following reviewers for their advice, which brought the quality of the manuscript to the highest level (in alphabetical order):

- M.ª Rosario Bautista Zambrana (*Universidad de Málaga*, Spain)
- Enrique Gutiérrez Rubio (*Palacký University Olomouc*, Czech Republic)
- Raquel Lázaro Gutiérrez (*Universidad de Alcalá de Henares*, Spain)
- María Araceli Losey León (*Universidad de Cádiz*, Spain)
- Elizabeth Marshman (*University of Ottawa*, Canada)
- Beatriz Martín Gascón (*Universidad Complutense de Madrid*, Spain)
- María Teresa Ortego Antón (*Universidad de Valladolid*, Spain)
- Janine Pimentel (*Universidade Federal do Rio de Janeiro*, Brazil)
- Jaime Sánchez Carnicer (*Universidad de Valladolid*, Spain)
- Jorge Soto Almela (*Universidad de Alicante*, Spain)

Table of Contents

Introduction 13

Míriam Buendía-Castro
1. Lexical domains in the field of adventure tourism 19

Gloria Cappelli
2. The language of accessible adventure tourism 35

Isabel Durán-Muñoz and Paula Prieto Mayo
3. Descriptive adjectives in adventure tourism: A corpus-assisted English-Spanish contrastive study 55

Patrick Goethals and Jasper Degraeuwe
4. Methodological advances in lexical pattern extraction: Examples from Spanish adventure tourism 85

Eduardo José Jacinto García
5. The argument structure of motion verbs in Spanish: A methodological proposal applied to *DicoAdventure* 111

Eva Lucía Jiménez-Navarro
6. Prepositional phrase collocations of motion verbs: A corpus-driven study in adventure tourism 139

Marie-Claude L'Homme
7. Frame Semantics and domain-specific resources 161

Elena Manca
8. Patterns and perspectives in the language of Italian and British walking holidays 185

Macarena Palma Gutiérrez
9. Syntactic alternations with verbs of motion: A corpus-driven analysis of the language of adventure tourism 211

Carmen Portero Muñoz
10. The use of compounds in the adventure tourism lexicon 229

Introduction

Adventure tourism is gaining importance in the tourism sector at present, since more and more people are becoming involved in sport, nature, and sustainability, and they are continuously looking for the active instead of the traditional holiday, like the well-known *sun and beach tourism*. Although passive activities related to traditional tourism still occupy a significant position in the global tourism economy, we are witnessing the widespread popularity of a range of alternative tourisms, such as the one with which we are concerned in this work, that is, adventure tourism.

This type of tourism offers numerous activities, from hiking to climbing or scuba diving, all of which take place in nature with different levels of difficulty and require an active involvement of tourists. In the same line, the language employed in this type of tourism is dynamic, full of motion and descriptions, and seeks to attract potential tourists by offering appealing adventure activities and natural spots. In the last years several works have been published that provide valuable insights into different linguistic, semantic, and pragmatic characteristics of this language (cf. Durán-Muñoz, 2014, 2019, 2021, 2022, 2024; Durán-Muñoz & Jiménez-Navarro, 2021, 2023; Durán-Muñoz & L'Homme, 2020; Jiménez-Navarro & Durán-Muñoz, 2024); however, there is still a gap in the literature with regard to a comprehensive description of its features.

The chapters selected for this volume give a detailed account of the language of adventure tourism and contribute to exploring this specialised domain from a linguistic, a semantic, and a pragmatic viewpoint. It includes 10 chapters dealing with different languages (Spanish, English, and Italian), either with a monolingual or a bilingual approach, and revolving around verbal units (Chapters 1, 5, and 9), adjectives (Chapter 3), phraseology (Chapter 6), compounding (Chapter 10), pragmatics (Chapter 8), and accessibility (Chapter 2). Moreover, the studies are corpus-based or corpus-driven – most of them employ the ADVENCOR corpus[1] – and implement different methodologies, from lexico-semantic

---

1 The ADVENCOR corpus is a bilingual (English-Spanish) specialised corpus about adventure tourism compiled in the framework of the R&D project "DicoAdventure: diseño y desarrollo de un recurso electrónico especializado bilingüe (inglés, español) sobre el turismo de aventura a partir de marcos semánticos" (Ref. UCO-1380857-F), co-funded by the Operational Programme FEDER 2014-2020 and

approaches (Chapters 1, 3, 5, 6, and 7) to natural language processing (Chapter 4). These enhance the understanding of current terminology and techniques.

In this context, the contributions gathered here conduct a substantially complete analysis of the main linguistic, semantic, and pragmatic features of the specialised language of adventure tourism, an ever-growing tourism subdomain, as well as a very up-to-date approach to corpus-assisted terminology studies. A brief description of them is provided next.

Chapter 1, by Míriam Buendía-Castro ("Lexical domains in the field of adventure tourism"), deals with verbal units. The author analyses the most representative 50 verbs in the ADVENCOR corpus and classifies them into domains and subdomains according to the Lexical Grammar Model (Faber & Mairal, 1999), which helps to acquire knowledge of this segment. Finally, she proposes a description of the subdomain *to feel sth good*, within the domain EMOTION, and proves that verbs belonging to the same subdomain exhibit similar semantic and syntactic behaviours.

Gloria Cappelli, in Chapter 2 ("The language of accessible adventure tourism"), presents the results of a qualitative, corpus-based analysis of the lexical features of accessible adventure tourism promotion. Her study is innovative and offers a novel approach to this field. She points out that this type of specialised discourse shows different characteristics with respect to both the discourse of general tourism and that of adventure tourism in terms of terminology and collocations. Additionally, she emphasises that much still needs to be done to ensure true inclusiveness in tourism discourse.

In Chapter 3 ("Descriptive adjectives in adventure tourism: A corpus-assisted English-Spanish contrastive study"), Isabel Durán-Muñoz and Paula Prieto Mayo present a corpus-assisted analysis of descriptive adjectives in the language of adventure tourism in both English and Spanish. These specialised units are analysed according to their meaning as either descriptive or evaluative, and grouped into a set of semantic categories established ad hoc for this research. Finally, differences and similarities between the two working languages are examined, and some relevant conclusions are drawn from the results.

Patrick Goethals and Jasper Degraeuwe offer a new approach to the language of adventure tourism in Chapter 4 ("Methodological advances in lexical pattern extraction: Examples from Spanish adventure tourism"). The authors describe several techniques (e.g., dependency parsing and semantic similarity calculation

---

the Consejería de Economía, Conocimiento, Empresas y Universidad of the Andalusian regional government, and led by Dr Isabel Durán-Muñoz.

based on non-contextual word embeddings and transformer-based language models) in order to prove how these recent innovations can contribute to a more fruitful use of mid-size corpora, such as the ADVENCOR corpus, in the domain under study and, therefore, improve extraction results to feed terminological resources like the *DicoAdventure* dictionary.[2]

Eduardo José Jacinto García, in Chapter 5 ("The argument structure of motion verbs in Spanish: A methodological proposal applied to *DicoAdventure*"), provides a thorough analysis and a classification of motion verbs in Spanish combining two different approaches: on the one hand, the Explanatory Combinatorial Lexicology (Mel'čuk et al., 1995; Mel'čuk & Milićević, 2020) and, on the other, Frame Semantics (Fillmore, 1976, 1982; Fillmore & Baker, 2010). According to the arguments of the analysed verbs, the author classifies them into verbs of displacement and verbs of manner of movement (Morimoto, 2001).

For her part, Eva Lucía Jiménez-Navarro focuses on the phraseological component of the language of adventure tourism in Chapter 6 ("Prepositional phrase collocations of motion verbs: A corpus-driven study in adventure tourism"). More specifically, she explores the prepositional phrases collocating with a set of motion verbs extracted from the ADVENCOR corpus. The findings are meaningful and show that collocations describing real motion are more common than those representing fictive motion. In addition, an analysis of the semantic roles expressed by the prepositions is also undertaken and the most recurrent prepositions are studied.

Chapter 7, by Marie-Claude L'Homme ("Frame Semantics and domain-specific resources"), deals with several questions that raise during the design and development of corpus-assisted specialised resources, such as what kinds of terms should be taken into consideration or what is the linguistic behaviour of terms. In this work, answers provided by Frame Semantics (Fillmore, 1976, 1982; Fillmore & Baker, 2010) and the lexical database FrameNet (Ruppenhofer et al., 2016) are examined as well as concrete implementations in two resources, namely *DicoAdventure*, a resource that records terms in the field of adventure tourism, and *DiCoEnviro*, a resource that contains environment terms.

---

2 http://olst.ling.umontreal.ca/dicoadventure/ [Last accessed: 28/09/2023]. This specialised resource has been designed and developed in the framework of the aforementioned R&D project "DicoAdventure: diseño y desarrollo de un recurso electrónico especializado bilingüe (inglés, español) sobre el turismo de aventura a partir de marcos semánticos" (Ref. UCO-1380857-F).

Elena Manca is the author of Chapter 8 ("Patterns and perspectives in the language of Italian and British walking holidays"), which contributes to the interdisciplinary and transdisciplinary research on adventure travel tourism by examining the discourse patterns used on websites promoting walking holidays in Italy and in the UK. This work explores the lexical, semantic, and pragmatic choices that characterise this type of discourse. Additionally, it aims to describe how authors and readers interact and how space is portrayed in relation to the participants following the theoretical models of metadiscourse (Hyland, 2005) and spatial description (Levinson, 1996; Taylor & Tversky, 1996). The results provide detailed information about the distinct features of promoting British and Italian walking holidays, which enable significant applications in the marketing domain.

Next, Macarena Palma-Gutiérrez is the author of Chapter 9 ("Syntactic alternations with verbs of motion: A corpus-driven analysis of the language of adventure tourism"). She analyses the syntactic and semantic characterisation of motion verbs, with a special focus on the structural and relational link between this type of verbs and their argument structure, as well as the syntactic alternations displayed by these verbs in the specialised domain of adventure tourism. Two subclasses of verbs of motion are examined: the subclass of "run" verbs (like *hike*) and the subclass of verbs that are vehicle names (like *canoe*). The main findings show that, although the intransitive basic or unmarked forms are frequent in the corpus examined, some verbs tend to occur more productively with other syntactically derived alternations.

Finally, Chapter 10, by Carmen Portero Muñoz ("The use of compounds in the adventure tourism lexicon"), explores the relevance of English compounding processes in the lexical repertoire of adventure tourism. Through the use of data extracted from ADVENCOR, this study aims to identify various compound patterns related to adventure tourism activities and their relative productivity. Initially, compounds are categorised based on their syntax and, subsequently, they are examined concerning the meanings of the first word, the head noun, and the internal association between their components, known as thematic relations. The research demonstrates that compounding plays a vital morphological role in creating new words within the segment under study.

With the focus on the language of adventure tourism, the contributions included in this volume implement fresh approaches and novel methodologies that represent an asset for linguists in general (terminologists, translators, interpreters, etc.) and encourage future lines of research, either in the domain of tourism or else. All chapters included have been written by renowned authors and peer-reviewed by experts, which ensures their overall quality and their important contribution to the field. We sincerely hope that this edited book becomes a valuable source of information for scholars, trainees, and professionals. Our ultimate goal is to raise awareness of the importance of analysing a specialised language (in this case, the language of adventure tourism) from different approaches (linguistic, semantic, and pragmatic), drawing attention to diverse grammatical categories (nouns, adjectives, verbs, etc.) so as to offer a glimpse of the need for designing and developing online specialised resources, such as *DicoAdventure*, and to contribute to the characterisation of this specialised discourse.

Isabel Durán-Muñoz and Eva Lucía Jiménez-Navarro

References

Durán-Muñoz, I. (2014). Aspectos pragmático-lingüísticos del discurso del turismo de aventura: Estudio de un caso. *Normas: revista de estudios lingüísticos hispánicos*, 4, 49–69. http://hdl.handle.net/10550/40573

Durán-Muñoz, I. (2019). Adjectives and their keyness. A corpus-based analysis in English tourism. *Corpora*, *14*(3), 351–378. https://doi.org/10.3366/cor.2019.0178

Durán-Muñoz, I. (2021). *DicoAdventure* y la terminología del turismo de aventura: Propuesta de diccionario en línea. In T. Barceló Martínez, I. Delgado Pugés, & F. García Luque (Eds.), *Tendencias actuales en traducción especializada, traducción audiovisual y accesibilidad* (pp. 395–417). Tirant Lo Blanch.

Durán-Muñoz, I. (2022). El trabajo terminográfico basado en corpus: El caso del recurso *DicoAdventure*. *Estudios de Traducción*, *12*, 109–118. https://doi.org/10.5209/estr.80584

Durán-Muñoz, I. (2024). Anglicismos en el turismo de aventura. In M. C. Balbuena Torezano (Ed.), *La traducción y la interpretación en tiempos de pandemia* (pp. 169–182). Peter Lang.

Durán-Muñoz, I., & Jiménez-Navarro, E. L. (2021). Colocaciones verbales en el turismo de aventura: Estudio contrastivo inglés-español. In G. Corpas Pastor, M.ª R. Bautista Zambrana, & C. M. Hidalgo-Ternero (Eds.), *Sistemas fraseológicos en contraste: Enfoques computacionales y de corpus* (pp. 121–142). Comares.

Durán-Muñoz, I., & Jiménez-Navarro, E. L. (2023). Motion verbs in adventure tourism: A lexico-semantic approach to fictive meaning. *IJES*, *23*(1), 27–48. https://doi.org/10.6018/ijes.532851

Durán-Muñoz, I., & L'Homme, M.-C. (2020). Diving into adventure tourism from a lexico-semantic approach: An analysis of English motion verbs. *Terminology*, *26*(1), 33–59. https://doi.org/10.1075/term.00041.dur.

Faber, P., & Mairal, R. (1999). *Constructing a lexicon of English verbs*. Mouton de Gruyter.

Fillmore, C. (1976). Frame Semantics and the nature of language. *Annals of the New York Academy of Sciences*, *280*(1), 20–32. https://doi.org/10.1111/j.1749-6632.1976.tb25467.x

Fillmore, C. (1982). Frame Semantics. In The Linguistic Society of Korea (Ed.), *Linguistics in the morning calm* (pp. 111–137). Hanshin.

Fillmore, C., & Baker, C. (2010). A frames approach to semantic analysis. In B. Heine & H. Narrog (Eds.), *The Oxford handbook of linguistic analysis* (pp. 313–340). Oxford University Press. https://doi.org/10.1093/oxfordhb/9780199544004.013.0013

Hyland, K. (2005). *Metadiscourse: Exploring interaction in writing*. Continuum.

Jiménez-Navarro, E. L., & Durán-Muñoz, I. (2024). Collocations of fictive motion verbs in adventure tourism: A corpus-based study of the English language. *Revista Española de Lingüística Aplicada* (online). https://doi.org/10.1075/resla.21042.jim

Levinson, S. (1996). Frames of reference and Molyneux's question: Cross-linguistic evidence. In P. Bloom, M. A. Peterson, L. Nadel, & M. F. Garrett (Eds.), *Space and language* (pp. 109–169). The MIT Press.

Mel'čuk, I., Clas, A., & Polguère, A. (1995). *Introduction à la lexicologie explicative et combinatoire*. Duculot.

Mel'čuk, I., & Milićević, J. (2020). *An advanced introduction to semantics. A meaning-text approach*. Cambridge University Press.

Morimoto, Y. (2001). *Los verbos de movimiento*. Visor Libros.

Ruppenhofer, J., Ellsworth, M., Petruck, M., Johnson, C., Baker, C., & Scheffczyk, J. (2016). *FrameNet II: Extended Theory and Practice*. https://framenet.icsi.berkeley.edu/fndrupal/index.php?q=the_book

Taylor, H. A., & Tversky, B. (1996). Perspective in spatial descriptions. *Journal of Memory and Language*, *35*, 371–391. https://doi.org/10.1006/jmla.1996.0021

Míriam Buendía-Castro

Universidad de Granada

mbuendia@ugr.es

1 Lexical domains in the field of adventure tourism

Abstract: In terminology, studies have traditionally focused on the study of nouns, often leaving aside other grammatical categories such as verbs or adjectives (Cabré, 1999; Rey, 1993 [1976]; Sager, 1990). However, verbs are one of the main lexical and syntactic categories of the language, so their study in specialised languages is essential. In this research, the 50 most representative verbs of the ADVENCOR corpus, a corpus specialised in adventure tourism, were analysed. These predicates were semantically classified according to the Lexical Grammar Model (Faber & Mairal, 1999) into domains and subdomains in order to discover the most prototypical ones in the field of tourism. This study shows, through the description of the subdomain *to feel sth good*, within the domain EMOTION, that verbs belonging to the same subdomain usually show very similar semantic and syntactic behaviour. This classification into lexical domains and subdomains can be particularly useful for the acquisition of knowledge in adventure tourism, as well as for the production and reception of specialised texts in this area.

Keywords: adventure tourism, corpus, lexical domain, meaning, verb

1. Introduction

In the first half of the twentieth century, most linguistic theories conceived of the combinatorial potential or subcategorisation of verbs mainly from a syntactic perspective. Both Structuralism (Saussure, 1990) and Generative Grammar (Chomsky, 1957, 1965) agreed that the study of the sentence and the meaning of words should be postponed and that sentences should be analysed first and foremost as purely syntactic structures. As Mairal and Faber (2005, p. 282) point out, linguistic theories that strive to account for syntactic structures and leave meaning for later inevitably run into problems because language is more about meaning than about grammatical constructions; they even go so far as to say that the syntax of lexical units depends on their meaning than the other way around, since meaning comes before syntax. Thus, as alternative paradigms to Formal Linguistics, in the late 1970s and early 1980s, some of the functional models were developed (e.g., the Lexical Grammar Model [Faber & Mairal, 1999] and the Role and Reference Grammar [Van Valin, 2005; Van Valin & LaPolla, 1997]) and functional-cognitive theories (e.g., Fillmore, 1977, 1982, 1985; Lakoff, 1987; Langacker, 1987, 1991). Functional and functional-cognitive theories agree on the idea that communicative situations determine meaning. Both approaches, therefore, account for the study of real language uses and variation. In other words, these theories are interested in language as a communicative phenomenon; in fact, the aim of functional theories is to describe language use in real communicative situations (Mairal et al., 2012, p. 222).

In contrast to what happens in Lexicology, generally speaking, Terminology has traditionally favoured the study of nominal units to the detriment of verbal units, despite the fact that verbs are essential categories for conveying meaning and, hence, their study in specialised languages is vital (Buendía-Castro, 2012, 2021a; Jiménez-Navarro, 2020). Thus, this chapter proposes a methodology for the extraction, description, and coding of verbs, illustrated within the specialised domain of adventure tourism. For this purpose, the 50 most representative verbs from the ADVENCOR corpus (Durán-Muñoz & Jiménez-Navarro, 2021) were analysed. They were semantically classified according to the Lexical Grammar Model (Faber & Mairal, 1999) into lexical domains and subdomains in order to discover the most prototypical ones in this field. This classification into lexical domains and subdomains can be especially useful for the acquisition of conceptual and linguistic information about adventure tourism.

This research is structured as follows. Firstly, Section 2 describes the main features of the Lexical Grammar Model. After that, Section 3 brings together the methodology and analysis of this study, and, more specifically, it shows the extraction of candidate verbs, the classification of predicates into domains and subdomains, as well as the linguistic realisations, conceptual categories, and semantic roles of arguments. Subsequently, Section 4 proposes a phraseological template model, and, finally, conclusions and future lines of research are listed in Section 5.

2. The Lexical Grammar Model: Domains and Subdomains

The Lexical Grammar Model (henceforth, LGM) organises onomasiologically the lexicon into semantic hierarchies that form lexical domains and subdomains. It focuses on verbs because much of our knowledge consists of events and states, most of which can be represented linguistically by verbs (Faber, 1999). Although the LGM was initially conceived for the general language, in recent years it has also been successfully applied to the study of specialised languages (Buendía-Castro, 2021b; Montero-Martínez & Buendía-Castro, 2017).

According to the LGM, the lexicon (i.e., the mental dictionary that every speaker of a language owns) is divided into 12 lexical domains. Each domain has one or two generic or superordinate verbs in terms of which all members of the domain are directly or indirectly defined. In this sense, the *genus* or core term of the definition of each lexeme marks the semantic territory covered by a specific domain or subdomain and is, thus, the factor that determines lexical domain membership. It is the semantic information of the *differentiae* that distinguishes one verb from another within the same domain. Table 1 shows the lexical domains and the superordinate verbs associated with each one (Faber & Mairal, 1999, p. 88):

Table 1. Lexical domains and superordinate verbs in the Lexical Grammar Model

Lexical domains	Superordinate verbs
EXISTENCE	*to be*
CHANGE	*to become different*
POSSESSION	*to have/give*
SPEECH	*to say*
EMOTION	*to feel*
ACTION	*to do/make*
MANIPULATION	*to use*
COGNITION/MENTAL PERCEPTION	*to know/think*
MOVEMENT	*to move (go/come)*
GENERAL PERCEPTION	*to become aware (notice/perceive)*
SENSE PERCEPTION	*to see/hear/taste/smell/touch*
POSITION	*to be/stay/put*

It is our assertion that each verb in the lexicon belongs to one of these categories or, in the case of polysemous verbs, they could even belong to different domains. Their membership depends on the most prototypically activated semantic features. In addition, lexical domains can be further subdivided into *subdomains*. Each subdomain focuses on a particular area of meaning and reflects a different specification of its content. This is especially important in terminology, where term meaning depends on the membership of terms in specialised knowledge subdomains.

The verbal lexicon of the LGM is mainly organised according to two axes: (1) the paradigmatic axis and (2) the syntagmatic axis. The *paradigmatic axis* codifies how elements are configured on the axis of selection by organising them onomasiologically in a hierarchy of domains and subdomains, and is also a determining factor in their syntax (Faber & Mairal, 1999, p. 80). As shown in Table 1, the initial lexical organisation in the LGM is based on definitional analysis. On the *syntagmatic axis*, each verb is associated with its inventory of complementation patterns or different activations of its argument structure. These patterns are instantiations of an underlying schema indicative of its quantitative and qualitative valency. The *quantitative valency* refers to the number of arguments, whereas the *qualitative valency* provides the semantic characterisation of arguments, that is, their semantic roles and selection constraints.

At this point, it is necessary to define what a semantic role is, since every linguistic theory that endeavours to account for verb meaning must also include a description of verb arguments (i.e., the participants) as well as how they relate to the predicate. *Semantic roles* generally express the set of properties that a verb entails for a given argument. Although almost every linguistic theory makes use of semantic roles in some form, there is considerable disagreement as to their number, nature, or function. In this research, the more general *thematic roles* of the Role and Reference Grammar (Van Valin, 2005; Van Valin & LaPolla, 1997) and the *argumentative roles* proposed by Goldberg in Construction Grammar (Goldberg, 1995, 2006) have been employed. Finally, certain roles necessary for the field of adventure tourism have been created and adapted, being two of the most recurrent ones EXPERIMENTER and STIMULUS. The EXPERIMENTER can be defined as the entity that sensory or emotionally receives a stimulus, whereas the STIMULUS is the entity that unintentionally produces a sensory or emotional sensation.

3. Methodology and Analysis

For the methodology and analysis of this study, the ADVENCOR corpus has been used, whose characteristics are described in Section 3.1. It should be noted that both arguments and predicates have been taken into account in this research. First, the extraction of candidate verbs (Section 3.1.) and the classification of predicates into lexical domains and subdomains (Section 3.2.) are detailed. Subsequently, the linguistic realisations and conceptual categories of the arguments are analysed, as well as the semantic functions they perform (Sections 3.3. and 3.4.).

3.1. Extraction of candidate verbs

The specialised corpus used in this study is the ADVENCOR corpus (Durán-Muñoz & Jiménez-Navarro, 2021). It contains recent original promotional texts in English about adventure tourism that were published on websites in English-speaking countries, such as the United Kingdom, the United States, and Ireland, and contains 1,064,664 words.

First, the most specialised verbs in the corpus were extracted with the terminology extractor TermoStat Web 3.0,[1] a freely available tool developed by Drouin (2003) at the University of Montreal (Canada). TermoStat identifies terms statistically by comparing the frequencies obtained in a specific and in a general corpus. The automatic analysis generated a total of 451 verbs. Figure 1 shows some of the candidate verbs through a lemmatised list (first column on the left) [*Candidat de regroupement*], followed by their frequency in the text [*Fréquence*], the score specificity assigned to each unit [*Score*], and, finally, the orthographic variants [*Variantes orthographiques*].

Candidat de regroupement	Fréquence	Score (Spécificité)	Variantes orthographiques	Matrice
climb	1933	101.46	climb climbs climbed climbing	Verbe
hike	1155	81.46	hike hikes hiking	Verbe
trek	925	73.28	trek treks trekked trekking	Verbe
enjoy	1322	67.39	enjoy enjoys enjoyed enjoying	Verbe
raft	753	66.16	raft rafted rafting	Verbe
jump	815	59.27	jump jumps jumped jumping	Verbe
offer	1514	54	offer offers offered offering	Verbe
skydive	495	53.8	skydive skydives skydiving	Verbe

Figure 1. Extraction of candidate verbs from the corpus (TermoStat)

---

1 http://termostat.ling.umontreal.ca [Last accessed: 28/09/2023].

As shown in Figure 1, some of the most representative verbs of the language used in adventure tourism are *climb, hike, trek, enjoy, raft, jump, offer,* or *skydive*. Although the results provided by TermoStat are quite accurate, it should not be forgotten that they are automatically obtained. Therefore, human evaluation is always necessary to determine whether the extracted terms are indeed terms whose meaning is relevant within the specific domain of study.

For this research, the first 50 verbs with the highest score of specialisation offered by TermoStat were studied. It should be clarified that this level of specificity does not correspond to the absolute frequency detected in the corpus. Thus, the 50 most specialised verbs provided by TermoStat and sorted in alphabetical order are the following: *abseil, access, ascend, book, camp, canoe, check, choose, climb, contact, cross, descend, dive, enjoy, exhilarate, experience, explore, fly, glide, guide, hike, include, jump, locate, love, navigate, offer, paddle, parachute, paraglide, participate, provide, raft, rappel, recommend, relax, require, ride, situate, skydive, soar, spend, surf, swim, travel, trek, visit, wade, walk,* and *zip*.

As shall be seen in Section 3.2., these verbs were classified into the lexical domains and subdomains proposed by the LGM on the basis of their definitions. This provided the most prominent lexical domains activated within the field of adventure tourism.

3.2. Classification of predicates into lexical domains and subdomains

Following the Lexical Grammar Model (Faber & Mairal, 1999), the selected verbs were classified into lexical domains and subdomains, according to the *genus* and *differentiae* of their definition. In addition, in order to profile the meaning of the predicates, their behaviour in the texts was studied, thanks to the Sketch Engine[2] software through the analysis of concordances in the ADVENCOR corpus. This greater semantic precision made it possible to divide each lexical domain into subdomains which focus on a more specific area of meaning, that is, the subdomains reflect the different specifications of the semantic content of the domain.

For instance, within the field of adventure tourism, the MOVEMENT domain is one of the most prototypical (Durán-Muñoz & L'Homme, 2020; Jacinto García, this volume; Palma Gutiérrez, this volume). Table 2 displays the subdomains (in bold italics) of the lexical domain MOVEMENT in adventure tourism. As shown, nine subdomains can be established: *to move on foot, to move by means of*

---

2 https://www.sketchengine.eu [Last accessed: 28/09/2023].

something, to move down through the air, to move down through a vertical surface, to move up, to move easily, to move through the water, to move across, to cause sth not to move for a short while. Each subdomain includes all the verbs activated in it (in italics) along with their definition. By way of illustration, within the subdomain *to move down through the air*, there are verbs such as *fall*, *skydive*, *parachute*, or *paraglide*, whose *genus*, directly or indirectly, corresponds to the superordinate of the hierarchy. Thus, *skydive* (*to jump from an aircraft and fall for as long as possible before opening a parachute*) is a hyponym of *fall* (*to move down through the air*), and *parachute* is a co-hyponym of *skydive*. Finally, *paraglide* is defined from *parachute* (*to parachute or to jump down from an elevated point such as a hill and to travel a long horizontal distance before landing*). The meaning will become more specialised as you descend the hierarchy, so *paraglide* will be more specialised than *fall*. Additionally, because of the transitivity of the hierarchy, hyponyms will inherit all the characteristics of the corresponding hypernyms.

It should be noted that the information extracted from both the corpus, through Sketch Engine, and from an English monolingual dictionary, the Cambridge Dictionary Online,[3] was extremely useful for elaborating the definitions.

Table 2. Subdomains of the lexical domain MOVEMENT activated in adventure tourism

MOVEMENT
to move on foot
walk: to move on foot
hike: to walk a long distance (especially in the countryside)
trek: to walk a long distance (especially over lands such as hills, mountains or forests)
paddle: to walk through water that is not very deep (especially at the edge of a beach)
to move by means of something
travel: to move from one place to another (by means of something)
canoe: to travel in a canoe
fly: to travel by aircraft
raft: to travel on a raft
ride: to travel in a vehicle (e.g., car, bus), or in a horse

(*continued*)

---

3 https://dictionary.cambridge.org/dictionary/english [Last accessed: 28/09/2023].

Table 2. Continued

MOVEMENT
surf: to travel on a wave, by means of a board
zip: to travel in a zipwire
to move down through the air
fall: to move down through the air
skydive: to jump from an aircraft and fall for as long as possible before opening a parachute
parachute: to jump from an aircraft using a parachute straight away
paraglide: to parachute or to jump down from an elevated point such as a hill and to travel a long horizontal distance before landing
to move down a vertical surface
descend: to move down
rappel: to descend a very steep slope by holding on to a rope that is fastened to the top of the slope (used mainly in the United States)
abseil: to descend a very steep slope by holding on to a rope that is fastened to the top of the slope (used mainly in the United Kingdom)
to move easily
glide: to move easily
to move up
ascend: to move up
climb: to ascend
to move through water
swim: to move through water by using your body
dive: to swim underwater
paddle: to move through water that is not very deep, especially at the edge of a beach, etc. (used mainly in the United Kingdom)
wade: to move through water that is not very deep, especially at the edge of a beach, etc. (used mainly in the United States)
to move across
cross: to move across
to cause sth not to move for a short while
stay: to cause sth not to move
camp: to stay for a short while

In addition to the domain MOVEMENT, previously mentioned, the analysis of the verbs retrieved from the corpus and their classification into lexical domains and subdomains revealed that, in general, the other most prototypical activated domains in the field of adventure tourism are ACTION, EMOTION, and EXISTENCE. This is because the concepts activated by adventure tourism are entities that bring an ACTION (e.g., *explore, visit, book, spend*) that activate, in most cases, a MOVEMENT (e.g., *climb, ascend, jump, skydive, parachute, paraglide*). All these events trigger an EMOTION (*feel, experience, enjoy, exhilarate, love, relax*) and, usually, everything happens in a place, hence the domain EXISTENCE is also very present (e.g., with verbs such as *situate* or *locate*) (Figure 2).

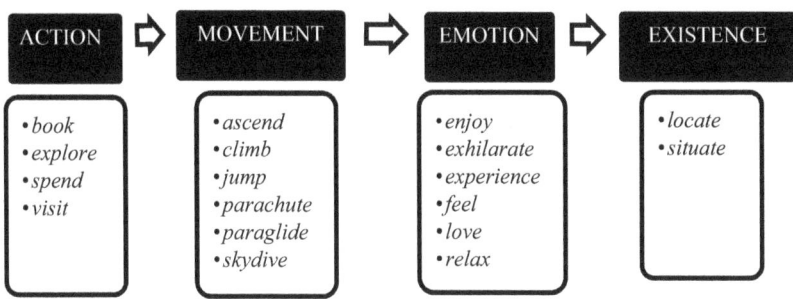

Figure 2. Main lexical domains activated in adventure tourism

3.3. Linguistic realisations and conceptual categories of the arguments

In addition to analysing the behaviour of predicates in the corpus, arguments were also studied. To this end, the linguistic realisations for the same argument were identified and semantic labels for the set of linguistic realisations that activate the same type of argument were assigned. *Linguistic realisations* of an argument are the instances of an argument in a corpus. When several linguistic realisations refer to the same argument, they designate the same type of entity, evoke the same type of conceptual structure, and exhibit similar semantic and syntactic behaviour, triggering the same type of semantic category. *Semantic categories* are generalisations of a set of terms that are assumed to have similar semantic and syntactic behaviour. For example, *T-shirt, sweatshirt, raincoat,* or *jumper* will be different linguistic realisations belonging to the semantic category of clothing. Finally, the semantic role activated by the arguments was specified. As previously mentioned, a *semantic role* describes the relationship between a verb and an argument. In this study, the more general roles of the Role and Reference Grammar (Van Valin, 2005; Van Valin & LaPolla, 1997) and the roles proposed by Goldberg in Construction Grammar (Goldberg, 1995, 2006) were used. In addition, certain roles necessary for the field of adventure tourism were created and adapted.

As an example, the arguments and semantic categories activated for the subdomain *to feel sth good*, within the domain EMOTION, are going to be described. Within this subdomain, we find verbs such as *enjoy, experience, feel,* or *love*, whose *genus*, directly or indirectly, corresponds to the superordinate of the hierarchy (Table 3):

Table 3. Verbs of the subdomain *to feel sth good*

to feel sth good
feel: to be aware of a particular emotion or sensation
experience: to feel sth
enjoy: to experience sth good
love: to enjoy sth very much

We now proceed to identify the linguistic realisations and semantic categories activated by these predicates through the study of the concordances of the corpus. Table 4 gives some examples taken from the ADVENCOR corpus:

Table 4. Linguistic realisations of the arguments for the subdomain *to feel sth good*

Everyone *loves* **extreme sports**.
Got **a bride to be** who *loves* the rush of **adrenaline**.
(…) so that **you** could *feel* the **excitement** and **enjoyment**.
Drivers to *feel* the **rush** that until now was only known in the racing world.
(…) where **you** will *enjoy* two hours of **tubing**.
You will be *enjoying* a tandem **sky dive**.
(…) as **you** *experience* the stunning **beaches**.
(…) as **you** *experience* this awesome **canyoning adventure**.

As shown in Table 4, it seems that two obligatory arguments are always required. The first one points to a human being which activates the semantic role of EXPERIMENTER[4] and is lexicalised through linguistic realisations

---

4 In *DicoAdventure* (http://olst.ling.umontreal.ca/dicoadventure/ [Last accessed: 28/09/2023]), the terminological resource on the specialised language of adventure tourism, this role is referred to as TOURIST.

such as *you* or *everyone*. The second obligatory argument suggests a positive emotion, a place, or an activity which activates the semantic role of STIMULUS and is lexicalised through lexical realisations such as *excitement* or *enjoyment*. These arguments can then be completed with attributes, such as *stunning* (*beaches*), or time or place complements (e.g., *two hours* (*of tubing*)). Table 5 provides a summary of the argument labelling in the subdomain *to feel sth good*.

Table 5. Argument labelling in the subdomain *to feel sth good*

Semantic role	EXPERIMENTER	*enjoy*	STIMULUS
Semantic category	HUMAN BEING	*experience* *feel* *love*	ACTIVITY, EMOTION, PLACE
Linguistic realisations	*you, bride to be, everyone*		*beach, enjoyment, excitement, extreme sport, tubing*

4. Phraseological Template

Once the verbs of each subdomain have been described, a phraseological template for each subdomain is proposed. In line with the previous example, we show in this section the phraseological template created for the subdomain *to feel sth good*, within the domain EMOTION. The aim is to collect all information regarding verbs and arguments by representing it in the form of requirements and restrictions, according to the linguistic realisations and semantic categories activated. Then, it is possible to establish phraseological generalisations for each lexical subdomain. Each phraseological template is composed of the following fields: (1) the name of the lexical domain, (2) the lexical subdomain, (3) the definition of the lexical subdomain, (4) a usage note, (5) the verbs activated in the subdomain along with their obligatory arguments (described in terms of their linguistic realisation in the corpus, the semantic category to which they refer, and the semantic role they play), and (6) examples of use drawn from the corpus, showing the verbs in context.

For example, according to the template shown in Table 6, when a predicate displays two obligatory arguments, being argument 1 a human being which activates the semantic role of EXPERIMENTER, and argument 2 a positive emotion, an activity, or place which activates the role of STIMULUS, the subdomain *to feel sth good* is activated. More specifically, the verbal predicates are *enjoy, experience, feel,* and *love.* Thus, all these combinations respond to a superordinate conceptual pattern, the phraseological pattern [EXPERIMENTER] feels (positive) [STIMULUS]. In turn, this combinatorial pattern takes place in the lexical domain EMOTION and in the subdomain *to feel sth good.* In addition, there are various less central features that may be activated in certain cases, such as the occurrence of semantic categories like PLACE, TIME, MANNER, and MEASURE.

Table 6. Phraseological template of the subdomain *to feel sth good*

Lexical domain: EMOTION			
Lexical subdomain: *to feel sth good*			
Definition: [EXPERIMENTER] feels (positive) [STIMULUS] or [THEME].			
Note: Additionally, the semantic categories of PLACE, TIME, MANNER, and MEASURE can be added.			
Semantic role	EXPERIMENTER	*enjoy*	STIMULUS
Semantic categories	HUMAN BEING	*experience* *feel*	ACTIVITY, EMOTION, PLACE
Linguistic Realizations	*you, bride to be, everyone*	*love*	*beach, enjoyment, excitement, extreme sport, tubing*
Usage contexts	Everyone *loves* **extreme sports**.		
	Got **a bride to be** who *loves* the rush of **adrenaline**.		
	(…) so that **you** could *feel* the **excitement** and **enjoyment**.		
	Drivers to *feel* the **rush** that until now was only known in the racing world.		
	(…) where **you** will *enjoy* two hours of **tubing**.		
	You will be *enjoying* a tandem **sky dive**.		
	(…) as **you** *experience* the stunning **beaches**.		
	(…) as **you** *experience* this awesome **canyoning adventure**.		

As shown, verbs restrict the possible arguments that can be combined with them (Buendía-Castro et al., 2014; Montero & Buendía-Castro, 2017). For instance, we observe that the verb *experience* needs a human being who feels an emotion. Likewise, emotions need to be experienced.

5. Conclusions

In this research, the 50 most representative verbs of the ADVENCOR corpus, specialised in adventure tourism, were analysed. For this purpose, it was necessary to study the behaviour of both nominal arguments and verbal predicates. Based on the classification of predicates into lexical domains and subdomains according to the LGM (Faber & Mairal, 1999), the arguments identified in our field of study, that is, adventure tourism, were classified and structured according to the semantic categories of the field, along with the semantic roles they activated. This enabled us to establish generalisations in the form of phraseological templates.

It was found that the most prototypical domains in this specialised language are ACTION, MOVEMENT, EMOTION, and EXISTENCE. It was also revealed, through the description of the subdomain *to feel sth good*, within the domain EMOTION, that verbs belonging to the same subdomain show remarkably similar semantic and syntactic behaviour. To put it differently, verbs belonging to the same subdomain usually exhibit the same number and type of arguments belonging to the same type of semantic categories and activate the same type of semantic roles. In line with the LGM, since all predicates in each subdomain are hierarchically organised according to their meaning, there is a focalisation of verb meaning as the hierarchy becomes more specific. In other words, verbs inherit semantic and syntactic patterns from their hypernym, but not all arguments of the hypernym will be activated by the hyponyms.

This method of coding information, also enriched with annotations on restrictions and language usage requirements, is particularly useful for knowledge acquisition about adventure tourism, as well as for the production and reception of specialised texts in this field. It can also be highly relevant for the collection of additional information that would enhance the development of the specialised resource *DicoAdventure*.

References

Buendía-Castro, M. (2012). Verb dynamics. *Terminology*, *18*(2), 149–166. https://doi.org/10.1075/term.18.2.01bue

Buendía-Castro, M. (2021a). *Verb collocations in dictionaries and corpus: An integrated approach for translation purposes*. Peter Lang.

Buendía-Castro, M. (2021b). *Adquisición de información conceptual y lingüística a través de los predicados y sus argumentos*. Comares.

Buendía-Castro, M., Montero-Martínez, S., & Faber, P. (2014). Verb collocations and phraseology in EcoLexicon. *Yearbook of Phraseology*, *5*(1), 57–94. https://doi.org/10.1515/phras-2014-0004

Cabré, M. T. (1999). *Terminology: Theory, methods and applications*. John Benjamins Publishing Company. https://doi.org/10.1075/tlrp.1

Chomsky, N. (1957). *Syntactic structures*. Janua Linguarum 4.

Chomsky, N. (1965). *Aspects of the theory of syntax.* MIT Press.

Drouin, P. (2003). Term extraction using non-technical corpora as a point of leverage. *Terminology, 9*(1), 99–117.

Durán-Muñoz, I., & Jiménez-Navarro, E. L. (2021). Colocaciones verbales en el turismo de aventura: Estudio contrastivo inglés-español. In G. Corpas Pastor, M.ª R. Bautista Zambrana, & C. M. Hidalgo-Ternero (Eds.), *Sistemas fraseológicos en contraste: Enfoques computacionales y de corpus* (pp. 121–142). Comares.

Durán-Muñoz, I., & L'Homme, M.-C. (2020). Diving into English motion verbs from a lexico-semantic approach: A corpus-based analysis on adventure tourism. *Terminology, 26*(1), 34–60. https://doi.org/10.1075/term.00041.dur

Faber, P. (1999). Conceptual analysis and knowledge acquisition in scientific translation. *Terminologie et Traduction, 2,* 97–123.

Faber, P., & Mairal, R. (1999). *Constructing a lexicon of English verbs.* Mouton de Gruyter.

Fillmore, C. J. (1977). Scenes and Frame Semantics. In A. Zampolli (Ed.), *Linguistic structures processing* (pp. 55–83). North Holland Publishing Company.

Fillmore, C. J. (1982). Frame Semantics. In The Linguistic Society of Korea (Ed.), *Linguistics in the morning calm* (pp. 111–137). Hanshin.

Fillmore, C. J. (1985). Frames and the semantics of understanding. *Quaderni Di Semantica, 6,* 222–254.

Goldberg, A. (1995). *Constructions. A construction grammar approach to argument structure.* University of Chicago Press.

Goldberg, A. (2006). *Constructions at work: The nature of generalization in language.* Oxford University Press.

Jacinto García, E. J. (This volume). The argument structure of motion verbs in Spanish: A methodological proposal applied to *DicoAdventure*. In I. Durán-Muñoz & E. L. Jiménez-Navarro (Eds.), *Exploring the language of adventure tourism: A corpus-assisted approach.* Peter Lang.

Jiménez-Navarro, E. L. (2020). *Treatment and representation of verb collocations in the specialised language of adventure tourism* [Doctoral dissertation, Universidad de Córdoba]. Helvia. https://helvia.uco.es/xmlui/handle/10396/20976

Lakoff, G. (1987). *Women, fire, and dangerous things.* University of Chicago Press.

Langacker, R. W. (1987). *Foundations of cognitive grammar: Theoretical prerequisites. Vol.1.* Stanford University Press.

Langacker, R. W. (1991). *Foundations of cognitive grammar: Descriptive application. Vol. 2.* Stanford University Press.

Mairal, R., & Faber, P. (2005). Decomposing semantic decomposition: Towards a semantic metalanguage in RRG. In *Proceedings of the 2005 International Conference on Role and Reference Grammar* (pp. 279–308). Academia Sinica.

Mairal, R., Peña Cervel, M. S., Cortés Rodríguez, F. J., & Ruiz de Mendoza Ibáñez, F. J. (2012). *Teoría lingüística: Métodos, herramientas y paradigmas* (segunda edición). Editorial Universitaria Ramón Areces.

Montero-Martínez, S., & Buendía-Castro, M. (2017). Clasificación semántica de colocaciones verbales para la adquisición y codificación de conocimiento experto: El caso de los riesgos naturales. *Revista Española de Lingüística Aplicada, 30*(1), 240–272. https://doi.org/10.1075/resla.30.1.10mon

Palma Gutiérrez, M. (This volume). Syntactic alternations with verbs of motion: A corpus-driven analysis of the language of adventure tourism. In I. Durán-Muñoz & E. L. Jiménez-Navarro (Eds.), *Exploring the language of adventure tourism: A corpus-assisted approach*. Peter Lang.

Rey, A. 1993 [1976]. *La terminologie: Noms et notions. Que sais-je?* Presses Universitaires de France.

Sager, J. C. 1990. *A practical course in terminology processing.* John Benjamins Publishing Company. https://doi.org/10.1075/z.44

Saussure, F. (1990). *Cours de linguistique générale.* Payot.

Van Valin, R. D. Jr. (2005). *The syntax-semantics-pragmatics interface: An introduction to role and reference grammar.* Cambridge University Press.

Van Valin, R. D. Jr., & LaPolla, R. (1997). *Syntax: Structure, meaning and function.* Cambridge University Press.

Acknowledgements

This work was supported by the Spanish Ministry of Science and Innovation under the Grant PID2020-118369GB-I00: "Transversal integration of culture and environmental terminological knowledge base", TRANSCULTURE.

Gloria Cappelli

University of Pisa
gloria.cappelli@unipi.it

2 The language of accessible adventure tourism

Abstract: This chapter presents the results of a qualitative, corpus-based analysis of the lexical features of accessible adventure tourism promotion. Accessibility in tourism aims to ensure equal opportunities for all tourists, including travellers with physical, psychological, or cognitive differences. Accessible adventure tourism is still relatively new and, given the nature of the leisure activities involved, the development of its promotional discourse is especially interesting, as it offers an optimal vantage point to investigate the way in which inclusion is (or is not) sufficiently warranted. The investigation of the most salient keywords suggests that this type of specialised discourse has distinctive features with respect to both mainstream general tourism and mainstream active tourism discourse. Moreover, the analysis of the collocational profile of terms used to represent adventure travellers with disabilities reveals that, although they are widely represented, the predominant rhetoric of promotion remains largely patronising. This suggests that much still needs to be done to ensure true inclusiveness in tourism discourse.

Keywords: accessibility, adventure tourism, disability, tourism discourse, vocabulary

1. Introduction

Adventure tourism is a fast-growing sector (Cheng et al., 2018; Durán-Muñoz & Prieto-Mayo, this volume; UNWTO, 2014), and there is no universal agreement on its definition. Rather than describing it in terms of the types of activities that fall under this "umbrella term", adventure tourism might be best described in terms of the dimensions involved in its practice (Janowski et al., 2021; Rantala et al., 2018). According to the Adventure Travel Trade Association (2011), adventure tourism includes at least two of these three aspects: interaction with nature, interaction with culture, and/or physical challenge. Activities have traditionally been divided into soft and hard, the former characterised by low-risk experiences and the need for basic skills, and the latter considered higher-risk activities requiring advanced skills and commitment (UNWTO, 2014). This classification has been criticised by more recent studies (cf. Janowski et al., 2021; Rantala et al., 2018), which have pointed out that its parameters can vary depending on many factors, including extrinsic and sometimes unpredictable conditions of the tourist experience (e.g., weather conditions, equipment failure) as well as intrinsic factors (e.g., psychological or physical barriers of tourists). Janowski et al. (2021) propose "a new and holistic conceptualisation of adventure tourism" (p. 9) in which the idea of soft and hard adventure is conceived as scalar.

Tourism has indeed been considered a universal right for over forty years, more specifically, since the UNWTO 1980 Manila Conference, which recognised it as "an aspect of the fulfilment of the human being" (UNWTO, 1980, p. 2). Efforts have been made in recent years towards the cultural normalisation of tourism, after much severe criticism of its marginalising and elitist nature (Gascón, 2016; Rubio-Escuderos et al., 2021). This revised view of physical challenge, skills, and risk-taking performance is reflected in the development of increasingly inclusive forms of tourism, capable of meeting the needs of all types of tourists, including those with disabilities who wish to experience adventure and active holidays. This change goes hand in hand with the development of a Social Model of Disability, which sees the barriers that reduce (and often impede) access to all domains of life as a consequence of the failure of society to adopt all the necessary measures to ensure the removal of such limitations (Barnes, 2019). This conceptualisation is echoed by the policies advocated by many institutional bodies, including the UN, which in its *Convention on the Rights of People of Disabilities* (2008) acknowledges their right to participate, like their non-disabled peers, in leisure, recreational, and tourist activities. A similar stance towards the promotion of accessibility and social inclusion is found in the UNWTO's *Recommendations on accessible information in tourism* (2016) and in much scholarly debate on the ethics of tourism (cf. Lovelock & Lovelock, 2013).

The interest in the globally growing market sector of travellers with disabilities (of many diverse aetiologies and types, either permanent or temporary) is therefore not surprising. Specialised companies are proposing a diversified offer of adventure tourism products, ranging from "soft skills" options, such as adaptive trekking and skiing, to "hard skills" options, such as adaptive bungee jumping or climbing. This has led to the publication of several socio-economic studies focusing on the present state as well as the opportunities and challenges of inclusive practices for the tourism sector and society at large. They promote the "development of Accessible Tourism for All" (Devile & Moura, 2015, p. 9) and stress the benefits of adventure tourism for people with disabilities as a way to improve social interaction and quality of life, as well as to foster personal development, strengthen self-esteem, and reduce feelings of vulnerability and lack of control over one's own life (Daniels et al., 2005; Devile & Moura, 2015; Goodwin et al., 2009; Mactavish et al., 2007; Moura et al., 2012). Other studies focus on the positive change brought about by technological progress and the availability of assistive technology devices (Zabłocki et al., 2022).

Discussing all the complex facets of the debate (including the criticism of the idea of "normalisation" and of a neutral and universal idea of both "normality" and "disability"; cf. Rubio-Escuderos et al., 2021) would exceed the limits of this chapter. Nevertheless, this brief overview should suffice to explain the growing interest in accessible and inclusive tourism discourse. To the best of our knowledge, however, no studies have yet been published that discuss the language adopted in materials produced to promote active tourism for people with special needs.

The present study addresses this gap in the literature. It presents the results of a corpus-based investigation of the language used in informative-promotional materials about adventure tourism products specifically designed for this market segment. There is indeed vast agreement on the idea that one of the ways in which inclusion is promoted is through communication (cf. Buhalis & Michopoulou, 2011; Cloquet et al., 2018; Eichhorn et al., 2008; Gandin, 2016, 2018, 2019; Gillovic et al., 2018). Without appropriate and accessible information, no participation is possible. As Gillovic et al. (2018) point out, "language has the power to create, condone or justify attitudes and behaviour" (p. 615). This analysis was therefore carried out with two main research questions in mind. Does the language of accessible adventure tourism have distinctive features compared to its mainstream counterpart? If so, does it reflect and contribute to promoting the principles of inclusiveness (Scheyvens & Biddulph, 2018)?

The chapter includes six sections. Section 2 presents previous research on communication and inclusion, accessible tourism, and active tourism discourse. Section 3 presents the corpus and the methodology used for the investigation, and Section 4 discusses some of the lexical features of accessible adventure tourism discourse as they emerge from the qualitative analysis of two sets of keywords. Section 5 offers some reflections on the inclusiveness of the discursive domain at issue based on the analysis presented in Section 4 and on the study of the collocational profile of the lexical items *person* and *people*. Finally, Section 6 discusses the study's limitations and draws some conclusions about accessible tourism communication and inclusiveness.

2. State of the Art

With the notable exception of Eichhorn et al. (2008), communication and language-related aspects of accessible tourism have remained largely unexplored until recently (Rubio-Escuderos et al., 2021). Several studies over the past decade have contributed to closing this gap, focusing either on the representation of travellers with disabilities in tourism discourse or on the importance of the availability and reliability of specialised information for inclusive tourism.

2.1. Accessibility, inclusiveness, and communication

Accessibility and inclusion are two related but distinct concepts. Cloquet et al. (2018) define inclusion "as a process toward a society in which every individual has an active role to play" (p. 222), and, in this sense, it depends on the provision of accessibility in all domains of human existence, including leisure and tourism. According to Gillovic and McIntosh (2020), this is an "aspirational ideal" (p. 9724), and inclusive tourism should address and reduce inequalities and foster understanding of the needs and rights of minorities. Scheyvens and Biddulph (2018) have proposed a model which includes seven elements which can be used to analyse tourism inclusiveness: (1) marginalised people as tourism producers and (2) as tourism consumers, (3) self-representation, (4) power relations, (5) the widening of the participation in decision-making, (6) the expansion of the tourism map, and (7) the promotion of mutual understanding and respect. Most of these factors involve communicative practices to some extent. Linguistic choices certainly contribute to and reveal aspects of self-representation, power relations, mutual respect, and understanding, and participation is only possible through knowledge, namely, if suitable information is made readily available and accessible (Blichfeldt & Nicolaisen, 2011).

In the case of tourism, poor communicative practices can become as limiting a barrier as any other physical obstacle (Eichhorn et al., 2008; Gandin, 2016; Lee et al., 2012). As Dann (1996) demonstrates, participating in tourism activities depends on information processing and affective choices, which determine travellers' decision-making in terms of the selection of products. The industry influences such processes through communicative choices aimed at creating the perfect and most persuasive match between the destination's pull factors and the potential clients' push factors (Cappelli, 2007).

For people with disabilities, finding all the necessary information about possible tourist experiences and destinations at the pre-trip stage is especially important. Eichhorn et al. (2008) define this as "instrumental" and "particularly crucial" (pp. 191–192) and claim that it is essential that all the actors involved in accessible tourism make every possible effort to eliminate "learned helplessness" (i.e., the belief in one's own innate inability to achieve goals as a result of enduring repeated aversive stimuli such as the difficulty to access information; cf. Lee et al., 2012; Seligman, 1972) for these travellers. However, although technological advancements and the wide availability of internet access and social media have contributed to the dissemination of information (Altinay et al., 2016; Buhalis & Michopoulou, 2011), several studies have shown that disability is still a taboo in general promotional tourism discourse (Gandin, 2016). Moreover, people with special needs are rarely represented. When they are, it is often through referential images focusing on special equipment or through reference to the availability of accessible facilities (e.g., bathrooms, ramps, lifts), sometimes resorting to terms that are no longer in use (e.g., *handicapped*) (Benjamin et al., 2021; Cloquet et al., 2018).

These communicative practices are not inconsequential. As mentioned in Section 1, it is widely recognised that linguistic choices contribute to creating the identity of social groups. They may perpetuate stereotypes and "also marginalise social and cultural groups" by creating invisible minorities (Benjamin et al., 2021, p. 308). It is probably for this reason that some studies have found that people with disabilities often do not trust the travel-related information produced by the industry and prefer resorting to different sources, such as user-generated content on blogs, social media, and sites publishing testimonials of first-hand experiences (Altinay et al., 2016; Buhalis & Michopoulou, 2011).

2.2. Accessible tourism discourse

Recent linguistic research has contributed useful insights into the characteristics of accessible tourism discourse. If Cloquet et al. (2018) and Benjamin et al. (2021) focus on the representation of people with disability in tourism marketing and promotional materials from a sociological perspective, Gandin (2016, 2018, 2019, 2021) presents a detailed corpus-based investigation of the linguistic features of this specialised domain, compares and contrasts it with "mainstream" tourism discourse, and discusses accessibility in English-to-Italian translations.

Gandin's contribution to the debate is especially interesting for the present discussion because it explores texts specifically designed to cater to travellers with disabilities. For this reason, it is reasonable to assume that they do not present the issues found in mainstream tourism discourse. Indeed, as could be expected, disabled travellers are well represented in her corpora through "lexical variation expressed by means of periphrases, synonyms, pronouns or metaphors employed to indicate disabled tourists and their specific impairments and concurrently reduce the negative implications of direct terms indicating disabled people and disabling pathologies" (Gandin, 2018, p. 69). The analysis of keywords reveals the predominance of three main thematic domains: types of disability, the level of physical accessibility of facilities and attractions, and the services available to disabled tourists. However, quite interestingly, Gandin (2016) points out that, even in these texts, the representation of people with disabilities is influenced by a medical view and that a certain hesitancy persists in referring to their condition.

Another interesting finding is the fact that accessible tourism discourse resorts only marginally to the typical strategies of tourism promotion (e.g., euphoria technique, conative language, linguistic creativity, and ego-targeting) and favours a more referential style aimed at providing objective and detailed information relative to the accessibility of the products promoted and the actual availability of specific services. Gandin (2018) also finds fewer descriptions of the actual destinations – which are usually characterised by the frequent use of very positive adjectives – and greater emphasis is placed on mobility and personal-care facilities. These distinctive features of accessible tourism discourse led the author to conclude that its primary aim is not to seduce and convince the potential tourist (as in most instances of promotional tourism discourse; cf. Cappelli, 2007; Maci, 2020; Manca, 2016, among others), but rather to inform and reassure disabled visitors on practical aspects of the trip.

2.3. Adventure tourism discourse

Accessible adventure tourism is a very specialised segment of the adventure tourism market, which has experienced rapid growth over the past decade. It is reasonable to assume that the discursive practices adopted to promote such niche products incorporate distinctive features that bring together the style of adventure tourism and accessible tourism discourse. For this reason, besides the research carried out on the language of the latter, a significant contribution to the present analysis has come from the studies of the former.

The pragmatic and lexical features of this type of discourse have been extensively investigated, both in English and Spanish, by Durán-Muñoz (2014, 2019), Durán-Muñoz and L'Homme (2020), and Estorell Pons (2016). These studies have shown that adventure tourism discourse shares many traits with non-specialised tourism discourse, for example, the reader is addressed directly – either via the second-person pronoun or imperative forms –, vocabulary is characterised by creativity and conciseness, and metaphors and other figures of speech frequently occur in the descriptions to attract tourists to the promoted destinations. Moreover, as in "general" tourism promotional materials, lexical items that convey very positive meanings are prevalent and contribute to these texts' predominantly persuasive and conative function.

However, some features appear to be distinctive. Thus, in adventure tourism discourse, the most salient keywords mainly relate to the activities offered, the participants, the equipment, and the locations. The latter are not described in the typical evocative terms used in texts which frame the destination as the main pull factor but instead in terms of their physical characteristics. These findings support Durán-Muñoz's (2014) observation that, in adventure tourism discourse, the referential function of texts appears more prominent than in other types of tourism discourse because great care is put into describing the activities and their settings.

In addition, Durán-Muñoz's (2019) investigation of adjective keyness reveals that the most representative adjectives qualify the tourist experience and feelings rather than places. Finally, Durán-Muñoz and L'Homme (2020) show that a crucial promotional role is played by verbs of motion, which confirms that one of the unique features of adventure tourism discourse is that potential tourists are targeted through a call to action, to move, to challenge one's own limits, and to reach goals (Durán-Muñoz, 2014) rather than via the construction of an alluring image of the destination.

3. Data and Methodology

To answer the research questions of whether the language of accessible adventure tourism has distinctive features compared to its mainstream counterpart and, if so, whether it reflects and contributes to promoting the principles of inclusiveness, a specialised 394,000-token corpus of authentic, informative-promotional texts advertising adventure tourism products for people with disabilities was created, which we called Accessible Active Tourism Corpus (henceforth, AATC). It includes texts published online on private companies' websites, English-language travel magazines, and blogs between 2000 and 2022, and it was compiled and queried via Sketch Engine.

Compared to the specialised corpus ADVENCOR (Durán-Muñoz & Jiménez-Navarro, 2021), AATC is more heterogeneous in terms of the text genres included. Since accessible active tourism is still a relatively small niche sector, the promotional material available online was still scant at the time of the collection. For this reason, other types of texts were added (i.e., product placement blog posts or travel magazine articles), which had been specifically written to promote the offer of specialised companies. This choice was made after a first qualitative analysis of the texts in the corpus revealed that one of the main features was the inclusion of testimonials written by happy customers. These testimonials seem to make up a significant portion of the commercial website component of the corpus, in line with what Altinay et al. (2016) and Buhalis and Michopoulou (2011) observed. It was, thus, assumed that blog posts and online magazine articles might be relevant sources of information and persuasion for people with disabilities exploring their options in terms of active tourism. Therefore, these types of entries were selected to include only instances that matched the testimonial text type. Nevertheless, despite the efforts required to sample the materials carefully, the choices made might have reduced the comparability of the data retrieved.

The first step of the analysis aimed to verify whether accessible active tourism discourse has distinctive lexical features. To this purpose, three lists of keywords were generated with the Sketch Engine terminology extraction function: a list of keywords for AATC (ListAAT01) and another one for ADVENCOR (ListAD01), both obtained by choosing the "Recreation" component of the English Web 2021 (enTenTen21) corpus as the reference general English corpus, since it also includes texts published online. The third list of keywords was retrieved for AATC using ADVENCOR as the reference corpus (ListAAT02). Only words occurring at least 10 times in the focus corpora were included.

The first 150 items on each list ordered by keyness were then selected for an in-depth examination. They were organised according to the part of speech, and nominal, verbal, and adjectival keywords were investigated separately. To verify some of the interpretive hypotheses, the study of adjectives was expanded to include all the items that occurred at least 10 times in AATC. Finally, the collocational profile of *person* and *people* was analysed in greater detail.

4. Lexical Features of Accessible Adventure Tourism

4.1. Keywords with respect to general English

AATC and ADVENCOR do not differ significantly in terms of the proportion of noun-, verb-, adjective-, and adverb-tokens they contain. However, if we compare the distribution of these parts of speech in the keyword lists, ListAAT01 and ListAD01 differ. Whereas, in both corpora, most keywords are nouns, more verbs feature in ListAD01, and more adjectives and adverbs are present in ListAAT01. This can possibly be interpreted as evidence of different forms of specialisation of accessible adventure tourism discourse with respect to mainstream adventure tourism when compared to general English.

As mentioned, the first 150 items on each list were selected for an in-depth analysis. Elements accidentally extracted (e.g., typos or HTML code fragments) were removed. Names of places or people comprised a large portion of each list: 30 % of the items included in ListAD01 and 40 % in ListAAT01, and they were all discarded as well. Reference to specific locations was expected, as destinations are an essential aspect of the tourist experience. However, in adventure tourism, locations seem to be especially important, in line with Durán-Muñoz and L'Homme's (2020) observation that they frequently co-occur with motion verbs and contribute to the description of the activities promoted in terms of relevant information about their setting. Interestingly, personal names are more frequent in AATC than in ADVENCOR, thus confirming the role of testimonials and first-hand experience reports in accessible adventure tourism promotion. The remaining keywords, organised in terms of parts of speech, were distributed in the two lists as indicated in Table 1.

Table 1. Distribution of keywords in POS categories (not including names)

	ListAD01	ListAAT01
Nouns	77 %	69 %
Verbs	17 %	7 %
Adjectives	6 %	22 %
Adverbs	--	2 %

4.1.1. Nouns in accessible adventure tourism discourse

Surprisingly, there is minimal overlap between the top noun keywords in ListAAT01 and ListAD01. Less than 30 % of the nouns are shared, all denoting activities (e.g., *sledding*). Most other nominal keywords denote different types (e.g., *glamping, safari, riding*), and some are very specialised (e.g., *paracanoe, hippotherapy*). This might simply mirror the specialisation of the accessible adventure tourism market and the offer available to people with disabilities.

Many keywords refer to equipment (e.g., *ski, canoe*), and most of the lexical items in this category are related to disability and adaptations (e.g., *sit-ski, outrigger, handcycle*). Others are used in a sense specific to accessible tourism discourse (e.g., *hoist*). The list also includes nouns denoting accessible facilities (e.g., *roll-in-showers, ramp*), types of accommodation (e.g., *glamphotel*), and places (e.g., *outlook, boardwalks*). *Wheelchair, accessibility,* and *disability* are among the top keywords too, and so are words referring to people, mainly denoting individuals practising an activity (e.g., *skier, equestrian*), but also specific categories of disabled people (e.g., *amputee, paraplegic*).

4.1.2. Verbs in accessible adventure tourism discourse

ListAAT01 includes only a few verbs, which refer to the activities that can be performed (e.g., *paddle, skydive, trek*). There is no significant qualitative mismatch between the two keyword lists, the only difference being the number of items included: ListAD01 includes 10 % more verbs than ListAAT01. This might be a significant difference in terms of promotional strategies. Since verbs have been described as distinctive of the communicative practices of adventure tourism discourse (Durán-Muñoz & L'Homme, 2020), the fact that only a few verbs have keyness in AATC with respect to general English might point towards a preference for different promotional strategies, focusing less on actions and motion and more on other aspects (e.g., description and/or evaluation).

4.1.3. Adjectives in accessible adventure tourism discourse

The category of adjectival keywords is perhaps the most interesting because it reveals noticeable differences between the accessible adventure tourism corpus and the mainstream adventure tourism corpus (cf. Durán-Muñoz & Prieto-Mayo, this volume). ListAAT01 includes more adjectives than ListAD01, but, whereas key adjectives in the latter are evaluative, as it is typical of tourism discourse and refer in very favourable terms to either the active experience (e.g., *thrilling*) or the tourist (e.g., *adventurous*), all the adjectives found in ListAAT01 are descriptive and refer for the most part to disability. Thus, we find adjectives describing places in terms of their accessibility (e.g., *accessible, barrier-free, paved, wheelchair-friendly*), activities (e.g., *adaptive, paralympic*), and people (e.g., *able-bodied, disabled, quadriplegic, poor-sighted*).

These results are not entirely unexpected, considering that the list was compiled against a corpus of general English, in which such terms certainly do not occur with the same frequency as in our specialised corpus. However, it supports Gandin's (2016) observation that the language of accessible tourism is prevalently referential and still portrays disability in the framework of the medical view.

To verify whether these data were influenced by the choice of focusing on the first 150 keywords only, the complete list of adjectives occurring at least 10 times in AATC was analysed and classified into descriptive and qualitative instances. Adjectives referring to quantity were excluded (e.g., *many, additional*). Some very positive qualifying adjectives were indeed found, as expected, since the texts pursue an informative-promotional aim (e.g., *amazing, beautiful, excellent*), although they only occur occasionally and are not varied. Nevertheless, the analysis confirmed that over 60 % of the adjectives used in the corpus are descriptive and refer to disability (e.g., *accessible, adaptive, blind, visual, medical*) and the accessibility and usability of the spaces and of the equipment (e.g., *all-terrain, smooth, manual, tailor-made*). Other adjectives reflect the level of difficulty of the activities (e.g., *easy, basic, appropriate, tricky, challenging*) or qualify people in relational terms (e.g., *friendly, keen, professional, supportive*). Overall, most items refer to places rather than the traveller's excitement and actions, as observed in Durán-Muñoz (2019). However, the tourist's experience seems to be quite relevant in accessible adventure tourism discourse too. Many adjectives, in fact, centre on the rewarding feelings of embarking on challenging activities (e.g., *worth, inspiring, memorable, unforgettable, unexpected, beneficial*) or on the tourist's state of mind before or after taking part in one (e.g., *emotional, nervous, confident*).

This analysis confirms the marked referential style of promotional materials for disabled adventure tourists, which is more evident than in mainstream adventure tourism discourse and supports the conclusions of Gandin's (2016, 2018, 2019, 2021) research on accessible tourism. At the same time, it shows that the tourist's experience is indeed an important part of promotional adventure tourism discourse. Nonetheless, whereas mainstream active travellers are persuaded through the prospect of a thrilling experience, the promotion for disabled active tourists relies on previous, equally abled customers' reassurance that services and facilities are indeed accessible and suitable, that the people involved are trustworthy, that the experience is rewarding, and that their possible doubts were shared by others before.

4.2. Keywords with respect to mainstream adventure tourism discourse

ListAAT02 includes the top 150 keywords in terms of keyness obtained by comparing AATC to ADVENCOR. The list was created to verify how accessible adventure tourism discourse differs from its general counterpart at the lexical level.

Only one verb (i.e., *paralyse*) figures among the top 150 keywords. Half the items on this list are names, which confirms the relevance of both locations and testimonials in accessible adventure tourism promotion. A few nouns that figure in ListAAT02 also figured in ListAAT01, such as some types of accommodation (e.g., *glamphotel*), activities (e.g., *game-drive*), and specialised equipment (e.g., *sit-ski*). This might be due to the different composition of the two corpora.

As could be expected, a wide range of nominal keywords unique to AATC refer to disability (e.g., *autism*) and people who are variously related to it (e.g., *therapist, carer, caregiver*). The fact that such a large number of nouns relate to accessibility and disability is somewhat unsurprising: it is to be expected that a specialised corpus such as AATC will contain these terms in a much higher proportion than a general English corpus and a mainstream adventure tourism corpus. It does, however, point to a difference in the relevant aspects for promoting adventure tourism to this market segment. Moreover, it confirms the "invisibility" of people with disabilities in general tourism discourse. The fact that 20 % of the distinctive keywords are adjectives related to disability and accessibility (e.g., *adaptive, barrier-free, assistive*) seems to confirm again that, as in the case of other types of accessible tourism discourse (Gandin, 2016, 2018), the promotional texts included in the focus corpus exploit the referential function more than other tourism texts.

5. Accessible Adventure Tourism Discourse and Inclusion

Despite being limited in scope, the overview of the lexical features presented in Section 4 can contribute to the debate on inclusive tourism. As mentioned in Section 2, Scheyvens and Biddulph (2018) point out that true inclusiveness can only be attained if progress is made in several components of accessible tourism, and one is the promotion of minorities as tourism producers. The corpus does not present people with disability as active tourism practitioners, in line with Scheyvens and Biddulph's (2018) observation that much still needs to be done in this sense. However, they are prominently featured as tourism consumers, both as addressees of promotion and as testimonials. The lexical choices made to represent disabled adventure tourists are strictly related to other relevant elements of inclusion as well, such as self-representation, the power relations with other participants in the experience, the participation in decision-making processes, and, as a consequence, the promotion of mutual understanding and respect.

Gillovic and McIntosh (2015, 2020) stress the importance of hearing the voices of accessible tourism stakeholders to make sure all parties are involved in decision-making. The predominance of terminology referring to the accessibility and suitability of facilities, places, and adaptive equipment can therefore be seen as the industry's attempt to facilitate decision-making by reducing learned helplessness, that is, by stating the facts and providing as much information as possible to prospective tourists with special needs. Moreover, accessible adventure tourism discourse seems to feature their voices prominently in the form of testimonials. However, surprise, relief, and reassurance are the most common elements conveyed by these voices, which reveal that, in fact, disabled adventure tourists do not yet have a real chance to shape the offer. Rather, they participate as passive recipients that "vet" the available products and make sure other fellow travellers with the same conditions know what to expect, as in examples (1) and (2):

(1) I was afraid to try any sports activities, more so with water sports. It was a surprise when I finally did it and I was so delighted that I did it with RMA. You guys were amazing! [AATC]
(2) Just a huge well done to you all! I am so grateful. [AATC]

Testimonials also allow tourists with disabilities to self-represent. The use of adjectives to describe emotions and feelings contributes to the characterisation of disabled tourists' experience of active tourism. They feel *emotional, glad, grateful, lucky, proud,* and *independent*. These lexical items reveal relief and joy in successfully participating in a challenging activity and overcoming barriers. Nevertheless, the widespread expression of gratitude reveals that what should be the recognition of a right seems to be perceived by these tourists themselves as a concession of able-bodied society.

Some interesting insights are offered by the agentive nouns in the two corpora. The wordlists of the two corpora were created with Sketch Engine. They were filtered to keep only nouns, and then the lists were manually scanned to keep only agent nouns. The relative frequency of *people* in AATC was three times higher than in ADVENCOR, once the Average Logarithmic Distance Frequency was controlled. Terms denoting people practising an activity (e.g., *climber, skier, glider*) were instead twice as frequent in the latter. Technical terms such as *customer* and *guest* are also more frequent in ADVENCOR, and so is *tourist*. In AATC, on the other hand, *traveller* is more frequently used, and so is *everyone*. These trends might be partly explained by the choice of recommended noun-first definitions for disabled individuals, rather than adjective-first. They might also evidence an attempt to be linguistically and conceptually inclusive, and a preference for more personal and empathetic communication for this target market segment.

A more in-depth analysis of the collocational patterns of *people* and *person* in AATC reveals some interesting facts. First, in the accessible adventure tourism corpus, there are very few adjectives referring to the qualities of individuals other than those describing types of impairments (e.g., *disabled, impaired, blind*). For instance, *adventurous* is only used in a handful of contexts, mainly to modify *activities* or *experience*, with the sole exception of one occurrence of *adventurous traveller*. When it occurs with *person*, it is used in a negative sentence in testimonials to support the idea that the activity is indeed suitable for all tourists:

(3) I am not the most adventurous person, but... [AATC]

In the vast majority of cases, *people* occurs in contexts describing types of disabilities. There is great lexical variation in this sense, and this is a positive feature of accessible adventure tourism discourse, which shows that the predominant view of disability in these texts is not monolithic as in much mainstream tourism discourse (Gandin, 2016, 2018), but rather it tries to convey the complexity and diversity of disability. Thus, we find an extensive range of variations with the pattern *people who have* (e.g., *adaptive needs, poor or no mobility, impairments*) and *people who have + past participle* (e.g., *experienced spinal cord injuries, suffered a stroke*). Other frequent lexical patterns with this function are *people who use* (e.g., *canes, walkers, wheelchairs*), *people who live with* (e.g., *disabilities, illness*), *people who suffer from* (e.g., *chronic back pain*), and *people with* (e.g., *disability, autism*). These forms are the most common and do not identify individuals with their disability, but the latter is described in terms of one of their attributes. Adjective-first appellatives (e.g., *blind people*) are, however, also used, and so is the pattern *people who are + disability* (e.g., *autistic, blind, deaf, impaired*), which reveals that much still needs to be done to promote the active participation of people with disabilities as tourism producers, and as producers of promotional tourism discourse.

Other interesting collocational patterns refer to what adventure travellers with disabilities presumably *want, seek,* and *look for.* These verbs are primarily used in the third person, so we can assume they are found in industry-generated promotional material rather than in testimonials. They collocate with expressions such as *to stay active, stress-free travel, wellness, partners,* mirroring the idea that inclusive tourism can offer both physical and psychological benefits and can favour social interaction. Interestingly, these people are also imagined as *looking for their next challenge,* which implies that they face other challenges in life and *work* to *overcome* them. Accordingly, in our AATC corpus, *people face barriers* and *issues*. For this reason, they supposedly *seek emotional healing* and *need a quiet, supportive environment* and *a more individualised pace.* From frequently occurring prepositional patterns such as *for + people* and *of + people*, a picture emerges of the ways in which the industry tries to meet these needs (and lure potential clients). The most common context in which we find these patterns is that of the activities (e.g., *riding for people with...*) and accommodation and facilities (e.g., *accessible wetrooms for people...*) offered, which are, for example, *adapted, modified, tailored, ideal* for them.

The industry also offers more intangible opportunities, such as *possibilities, choice, inclusion,* and *fun,* and all of the above in a *welcoming space,* even a *haven,* through *consideration* and *paying attention* to their needs. Moreover, they wish to listen to the *stories* and the *voice of people* with disabilities to ensure their *participation* and *well-being*. As is evident, these lexical choices try to match the accessible and inclusive tourism agenda as much as possible and, through the use of empathetic language, they try to persuade potential clients, who are apparently thankful for this willingness to help.

If we look at the patterns in which *people* and *person* occur in object position, it is evident that, despite what might be an honest way to pursue inclusiveness, the rhetoric exploited is quite patronising. The industry promises to welcome and accommodate tourists with special needs, but, more frequently, it offers to *assist* and *support* them. They will *get* them *out* or *into the water* or *into a swimsuit*, or simply they will *get* them *excited*. They will *encourage* them to practise adventure activities they would not have thought were possible and will *allow* them to do so through *adaptive strategies* and *assistive technologies*. They will also take care of their psychological well-being by *enabling* them to *experience joy*, *feel happy and more confident*, *achieve their goals*, and *take an active role* in their holiday experience. They will *give* them *opportunities*, *chances*, *ideas* for things to do, and the *time* to do them *at their own pace*, thus *empowering* them.

As can be noted, this representation of people with disability is predominantly ableist: their needs and desires are interpreted from the perspective of the able-bodied producers and what they believe adventure tourists with disabilities should want, value, and aspire to. A prevalently medical view of disability emerges, which sees the industry as an entity that will work around physical impairments to involve these tourists in an otherwise able-bodied tourist domain.

These communicative choices reveal non-symmetrical power relations among stakeholders, which unfortunately also surface in the feelings of surprise and gratitude in the testimonials chosen to promote the tourism products advertised as if companies were making concessions and people with disabilities should recognise and appreciate it. In conclusion, it is not surprising, then, that in some of the comments to product placement articles and blog posts – which were not included in the corpus – a voice of dissent towards this attitude can be heard, as exemplified by example (4):

(4) You constantly hear, "Oh, excuse me. I'm sorry, I'm sorry. Oh, come here, I'll help you." It's out of the kindness of people's hearts but it gets overwhelming and it's not what we need. [AATC]

6. Conclusions

The analysis of the lexical features of AATC suggests that accessible tourism discourse can be seen as a highly specialised type of tourism discourse sharing characteristics of both general accessible and general adventure tourism communication. It confirms the distinctive preference for referentiality rather than the typical persuasive style of general tourism discourse, which is evidenced by the choice of largely descriptive adjectives and nouns that refer to accessibility and practical aspects relevant to travellers with special needs. In this sense, we can reasonably assume that this atypical preponderance of referential information is a distinctive communicative strategy for promoting inclusive adventure tourism products, as it might serve the function of avoiding disabled travellers' learned helplessness, thus favouring decision-making and promoting inclusion.

In addition, the analysis confirms rhetorical emphasis on pull factors also exploited by mainstream adventure tourism discourse, that is, the descriptions of activities and of the emotions and feelings produced by partaking in them. Contrary to general adventure tourism, though, accessible adventure tourism offers a more "intimist" version of such descriptions and insists on feelings of insecurity (prior to embarking on the activities) as opposed to feelings of fulfilment and pride as a reward for accepting the challenge.

This contributes to the self-representation of adventure tourists with a disability, who are widely represented in the corpus through a great variety of expressions. This indicates that, as could be expected, the accessible adventure tourism discourse does not see disability as a monolithic entity and recognises the complexity of the population it targets. At the same time, the study has revealed that the invisibility of people with special needs is still an issue in mainstream adventure tourism discourse.

Much still needs to be done to make tourism communication truly inclusive, including accessible promotional tourism discourse. If the voices of people with disability are, in fact, heard in the testimonials widely used as a promotional strategy in AATC, such voices often convey expressions of gratitude and surprise, which contributes to the generalised ableist perspective adopted by the promotional texts in the corpus. Access and inclusion should indeed be accepted as a right and not perceived as a concession.

This power imbalance between the industry, which offers to share products otherwise designed for non-disabled travellers, and the potential clients, who thankfully accept being welcomed into this specialised sector, also emerges in the representation of people with disabilities proposed in the promotional texts. The picture that is verbally drawn is that of adventure tourism providers on a mission to *help* passive people with disabilities *take the reins* of their existence, to *enable* them to take an active role, and to be independent. We are told that there is a community of welcoming tourism practitioners ready to *inspire, support,* and *explore what is possible* to *get people with special needs involved.*

In this rhetoric, *helping people with disabilities* becomes *doing good*, and saying that a person with special needs *will be able to* do something becomes a synonym for *having the ability* to participate in a certain activity rather than *having the opportunity* to, which is the case in the corpus of general adventure tourism (ADVENCOR).

As for the limitations of this study, it has the ones of most qualitative analyses. For instance, it focuses only on a selection of lexical items and their occurrences, so it would certainly be interesting to conduct a more extensive and systematic investigation of the most salient lexical items and their collocational profiles (e.g., nouns describing stakeholders in adventure tourism). Other interesting insights might emerge from investigating modality and evaluative language in the AATC corpus. Finally, it would be helpful to compare both specialised adventure tourism discourse corpora to a general tourism discourse corpus.

Despite these shortcomings, we believe that these preliminary observations show that much still needs to be done to make the rhetoric of accessible tourism promotion truly inclusive so that tourism discourse can contribute to promoting participation, improving mutual understanding and respect, and ultimately creating a fairer society for all.

References

Adventure Travel Trade Association, The George Washington University, & Vital Wave Consulting. (2011). *Adventure Tourism Development Index 2010 Report.* https://www.adventureindex.travel/docs/atdi_2010_report.pdf

Altinay, Z., Saner, T., Bahçelerli, N. M., & Altinay, F. (2016). The role of social media tools: Accessible tourism for disabled citizens. *Journal of Educational Technology & Society, 19*(1), 89–99.

Barnes, C. (2019). Understanding the social model of disability: Past, present and future. In N. Watson, A. Roulstone, & C. Thomas (Eds.), *Routledge handbook of disability studies* (pp. 14–31). Routledge.

Benjamin, S., Bottone, E., & Lee, M. (2021). Beyond accessibility: Exploring the representation of people with disabilities in tourism promotional materials. *Journal of Sustainable Tourism, 29*(2–3), 295–313.

Blichfeldt, B. S., & Nicolaisen, J. (2011). Disabled travel: Not easy, but doable. *Current Issues in Tourism, 14*(1), 79–102.

Buhalis, D., & Michopoulou, E. (2011). Information-enabled tourism destination marketing: Addressing the accessibility market. *Current Issues in Tourism*, *14*(2), 145–168.

Cappelli, G. (2007). *Sun, sea, sex and the unspoilt countryside. How the English language makes tourists out of readers*. Pari Publishing.

Cheng, M., Edwards, D., Darcy, S., & Redfern, K. (2018). A tri-method approach to a review of adventure tourism literature: Bibliometric analysis, content analysis, and a quantitative systematic literature review. *Journal of Hospitality & Tourism Research*, *42*(6), 997–1020.

Cloquet, I., Palomino, M., Shaw, G., Stephen, G., & Taylor, T. (2018). Disability, social inclusion and the marketing of tourist attractions. *Journal of Sustainable Tourism*, *26*(2), 221–237.

Dann, G. M. (1996). *The language of tourism: A sociolinguistic perspective*. Cab International.

Daniels, M., Rodgers, E., & Wiggins, B. (2005). "Travel Tales": An interpretive analysis of constraints and negotiations to pleasure travel as experienced by persons with disabilities. *Tourism Manegement*, *26*, 919–930.

Devile, E. L., & Moura, A. (2015). Adventure tourism for people with disabilities in Portugal: Opportunities and challenges. In R. Melo, R. Mendes, A. S. Damásio, & A. Ramos (Eds.), *Sport tourism: New challenges in a globalized world* (pp. 9–21). Coimbra College of Education.

Durán-Muñoz, I. (2014). Aspectos pragmático-lingüísticos del discurso del turismo de aventura: Estudio de un caso. *Normas: revista de estudios lingüísticos hispánicos*, *4*, 49–69.

Durán-Muñoz, I. (2019). Adjectives and their keyness. A corpus-based analysis of tourism discourse in English. *Corpora*, *14*(3), 351–378. https://doi.org/10.3366/cor.2019.0178

Durán-Muñoz, I., & Jiménez-Navarro, E. L. (2021). Colocaciones verbales en el turismo de aventura: Estudio contrastivo inglés-español. In G. Corpas Pastor, M.ª R. Bautista Zambrana, & C. M. Hidalgo-Ternero (Eds.), *Sistemas fraseológicos en contraste: Enfoques computacionales y de corpus* (pp. 121–142). Comares.

Durán-Muñoz, I., & L'Homme, M.-C. (2020). Diving into adventure tourism from a lexico-semantic approach: An analysis of English motion verbs. *Terminology*, *26*(1), 33–59. https/doi.org/10.1075/term.00041.dur

Durán-Muñoz, I., & Prieto-Mayo, P. (This volume). Descriptive adjectives in adventure tourism: A corpus-assisted English-Spanish contrastive study. In I. Durán-Muñoz & E. L. Jiménez-Navarro (Eds.), *Exploring the language of adventure tourism: A corpus-assisted approach*. Peter Lang.

Eichhorn, V., Miller, G., Michopoulou, E., & Buhalis, D. (2008). Enabling access to tourism through information schemes? *Annals of Tourism Research*, *35*(1), 189–210. https://doi.org/10.1016/j.annals.2007.07.005

Estorell Pons, M. (2016). Codificación y variación léxica en el turismo activo: Análisis de un corpus de textos electrónicos promocionales y normativos. In M. López Santiago & D. Giménez Folqués (Eds.), *El léxico del discurso turístico 2.0* (pp. 61–85). Institut Universitari De Llengües Aplicades Modernes (IULMA).

Gandin, S. (2016). Il linguaggio e la traduzione del turismo accessibile. Uno studio preliminare. *Lingue e Linguaggi*, *18*, 47–64. https://doi.org/10.1285/i22390359v18p47

Gandin, S. (2018). Tourism promotion and disability: Still a (linguistic) taboo? A preliminary study. In M. Bielenia Grajewska & M. E. Cortès de los Rìos (Eds.), *Innovative perspectives on tourism discourse* (pp. 55–73). IGI Global.

Gandin, S. (2019). L'accessibilità nel linguaggio del turismo accessibile: Riflessioni linguistiche, contestuali e traduttive. *Altre Modernità: Rivista di Studi Letterari e Culturali*, *21*, 33–54. https://doi.org/10.13130/2035-7680/11631

Gandin, S. (2021). Accessible tourism discourse and the pandemic. Linguistic resilience and communicative strategies in the promotion of *tourism for all*. *Iperstoria*, *0*(18), 10–32. https://doi.org/10.13136/2281-4582/2021.i18.1044

Gascón, J. (2016). Deconstruyendo el derecho al turismo/Deconstructing the right to tourism. *Revista CIDOB d'Afers Internacionals*, *113*, 51–69.

Gillovic, B., & McIntosh, A. (2015). Stakeholder perspectives of the future of accessible tourism in New Zealand. *Journal of Tourism Futures*, *1*(3), 223–239.

Gillovic, B., & McIntosh, A. (2020). Accessibility and inclusive tourism development: Current state and tuture agenda. *Sustainability*, *12*(22), 9722. https://doi.org/10.3390/su12229722

Gillovic, B., McIntosh, A., Darcy, S., & Cockburn-Wootten, C. (2018). Enabling the language of accessible tourism. *Journal of Sustainable Tourism*, *26*(4), 615–630. https://doi.org/10.1080/09669582.2017.1377209

Goodwin, D., Peco, J., & Ginther, N. (2009). Hiking excursions for persons with disabilities: Experiences of interdependence. *Therapeutic Recreation Journal*, *41*(4), 298–325.

Janowski, I., Gardiner, S., & Kwek, A. (2021). Dimensions of adventure tourism. *Tourism Management Perspectives*, *37*, 100776. https://doi.org/10.1016/j.tmp.2020.100776

Lee, B. K., Agarwal, S., & Kim, H. J. (2012). Influences of travel constraints on the people with disabilities' intention to travel: An application of Seligman's helplessness theory. *Tourism Management*, *33*(3), 569–579.

Lovelock, B., & Lovelock, K. M. (2013). *The ethics of tourism: Critical and applied perspectives*. Routledge.

Maci, S. M. (2020). *English tourism discourse: Insights into the professional, promotional and digital language of tourism*. Hoepli Editore.

Mactavish, J., MacKay, K., Iwasaki, Y., & Betleridge, D. (2007). Family caregivers of individuals with intellectual disability: Perspectives on life quality and the role of vacations. *Journal of Leisure Research, 39*(1), 127–155.

Manca, E. (2016). *Persuasion in tourism discourse: Methodologies and models*. Cambridge Scholars Publishing.

Moura, A., Kastenholz, E., & Pereira, A. (2012). Aliviar o stress de indivíduos com incapacidade: O potencial da prática turística. *Revista Turismo & Desenvolvimento, 17/18*(3), 1387–1401.

Rantala, O., Rokenes, A., & Valkonen, J. (2018). Is adventure tourism a coherent concept? A review of research approaches on adventure tourism. *Annals of Leisure Research, 21*(5), 539–552. https://doi.org/10.1080/11745 398.2016.1250647

Rubio-Escuderos, L., García-Andreu, H., & Ullán de la Rosa, J. (2021). Accessible tourism: Origins, state of the art and future lines of research. *European Journal of Tourism Research, 28*, 2803. https://doi.org/10.54055/ejtr.v28i.2237

Scheyvens, R., & Biddulph, R. (2018). Inclusive tourism development. *Tourism Geographies, 20*(4), 589–609. https://doi.org/10.1080/14616688.2017.1381985

Seligman, M. E. (1972). Learned helplessness. *Annual Review of Medicine, 23*(1), 407–412.

UNWTO (1980). Manila Declaration on World Tourism. *UNWTO Declarations*, 1(1), 1–34. https://www.e-unwto.org/doi/pdf/10.18111/unwtodeclarati ons.1980.01.01

UNWTO (2014). *Global Report on Adventure Tourism*. https://doi.org/10.18111/ 9789284416622

UNWTO (2016). *Recommendations on Accessible information in Tourism*. UNWTO. https://doi.org/10.18111/9789284417896

Zabłocki, M., Branowski, B., Kurczewski, P., Gabryelski, J., & Sydor, M. (2022). Designing innovative assistive technology devices for tourism. *International Journal of Environmental Research and Public Health, 19*(21), 14186. https://doi.org/10.3390/ijerph192114186

Isabel Durán-Muñoz

Universidad de Córdoba
iduran@uco.es

Paula Prieto Mayo

Universidad de Córdoba
l82prmap@uco.es

3 Descriptive adjectives in adventure tourism: A corpus-assisted English-Spanish contrastive study

Abstract: This chapter attempts to shed some light on the importance of adjectives in the linguistic characterisation of a specialised language and, more concretely, of the language of adventure tourism in both English and Spanish. It seeks to understand the role that adjectives play in this specific subdomain. To do so, this chapter approaches the adjectives extracted from a representative specialised bilingual (English, Spanish) comparable corpus, ADVENCOR, with the purpose of analysing them, classifying them according to their meaning as either descriptive or evaluative, and categorising the set of descriptive adjectives into 14 semantic categories established ad hoc for this research. In addition, differences and similarities between the two working languages (English and Spanish) are examined, and some relevant conclusions are drawn from the results. In broad terms, it is noticeable that the use and function of adjectives are rather different in each of these languages in this tourism subdomain. English leans to use adjectives more frequently, being most of them evaluative, in contrast to Spanish, which prefers the use of descriptive adjectives in its texts.

Keywords: adjectivisation, adventure tourism, contrastive study, corpus-assisted study, semantic meaning

1. Introduction

Adventure tourism has been rapidly growing over the last two decades and is forecasted to continue to expand (UNWTO, 2014, p. 10) since more and more travellers seek active authentic experiences combined with natural and sustainable values. This type of tourism consists in the practice of adventure activities, such as climbing, rafting, or canyoning, which take place outdoors in nature with greater or lesser intensity and risk. Although activities related to traditional tourism like sun and beach, that is, with a more passive participation by the tourist, still occupy a significant position in the global tourism economy, it is worth noting the rising popularity of adventure tourism, along with other types of alternative tourism, such as ecotourism or nature tourism.

As mentioned in Cappelli (this volume), defining adventure tourism has become extremely difficult, mainly due to the subjective nature of the concept of *adventure* and the overlap with other alternative forms of tourism (Buckley, 2006, 2010; Swarbrooke et al., 2003). Despite this, there have been numerous attempts to define it. Some of them[1] emphasise physical aspects, such as activity, wilderness, or equipment (cf. Buckley, 2006; Hudson, 2003; Lee et al., 2015; Sung et al., 2000), whereas others place the focus on psychological aspects of the adventure experience, like excitement, challenge, or self-development (cf. Muller & Cleaver, 2000; Swarbrooke et al., 2003). Moreover, the concept of risk, to a greater or lesser extent, is often included as a fundamental part of this type of tourism (Page et al., 2005). By way of example, the US-based Adventure Travel Trade Association (ATTA, 2022) defines *adventure tourism* as a "type of tourism involving exploration or travel with perceived (and possibly actual) risk, and potentially requiring specialized skills and physical exertion". As a consequence, we can state that adventure tourists play an active role in these activities with a high degree of involvement, demand a real adventure experience in nature, and search for certain elements of risk and challenge.

From a linguistic perspective, promotional texts about adventure tourism are characterised by different features, like the use of synonyms and quasi-synonyms, cultural terms, foreign words, and motion verbs, among others (Durán-Muñoz, 2014; Durán-Muñoz & L'Homme, 2020). Another relevant aspect of this specialised language is the widespread use of adjectives, either evaluative or descriptive, as they are employed as a way to persuade and attract potential tourists (cf. Durán-Muñoz, 2019; Edo Marzá, 2011). Adjectives play an important part in creating destination image and are understudied in this field for two main reasons: on the one hand, the language of tourism has recently been considered worth studying as a domain-specific discourse, as it has usually been regarded as part of the general language; and, on the other, adjectives are not traditionally identified as terms from a terminological viewpoint (cf. Section 2) and, therefore, are overlooked in general and not only in the tourism sector. Furthermore, there is a wide range of textual genres and tourism subdomains

---

1 A comprehensive conceptualisation of adventure tourism is provided in Janowski and Reichenberger (2019).

(adventure, nature, sun and beach, wellness, etc.) and their respective features, including adjectivisation, are still poorly covered by research.

To contribute to filling in this gap, this chapter focuses on the adjectivisation in a specific tourism subdomain, adventure tourism, and carries out a corpus-assisted contrastive study (English and Spanish) with the aim of providing some insights into the usage of adjectives in this specialised language as well as their main semantic features. More specifically, this research aims to investigate this specific aspect by answering the following questions: (1) which type of adjectives, that is, descriptive or evaluative, is more frequent in this subdomain, (2) what semantic categories are more frequent among the descriptive adjectives in the corpus, and (3) what the main differences in their usage in the languages under study are. To do so, the chapter is structured as follows. Section 2 reviews previous research on adjectivisation in the language of tourism and semantic classifications of adjectives, as well as provides some knowledge about the importance of adjectives in the language of adventure tourism. Next, Section 3 centres on the methodology followed to carry out this study by describing the different steps: corpus compilation, extraction of candidates and selection, and analysis and classification of selected adjectives in both languages. Section 4 gives an account of the main results and findings of the study, and, finally, Section 5 presents the concluding remarks and discusses future lines of research.

2. Adjectivisation and the Language of Tourism

Although outnumbered by nouns and verbs (Leech, 1989), adjectives are a significant open-word class in both English and Spanish and, from a grammatical and semantic perspective, they hold the same degree of importance as the other content words in the linguistic code. However, terminographers have traditionally considered terms only in noun form (cf. L'Homme, 2002), since these units are regarded as conveyors of meaning and, thus, labels for concepts (cf. ISO 704, 2009, pp. 23–24). This means that adjectives, adverbs, and verbs have been presented as modifiers of nouns or noun derivatives and, therefore, have remained under-studied units within the framework of specialised languages. Fortunately, more and more studies claim that content units different from nouns – that is, verbs, adjectives, and adverbs – are relevant to the understanding of how knowledge is conveyed in specialised texts and that they should be considered terms. In fact, previous research (Alonso & Torner, 2010; Dancette & L'Homme, 2004; Durán-Muñoz & L'Homme, 2020; L'Homme, 2002; Lorente, 2001; Pitkänen-Keikkilä, 2015) has proved that not only nouns but also verbs and adjectives can play a significant role in specialised languages by characterising an entity or indicating essential features of it. As such, they should be included in specialised resources. Moreover, to fully cover the linguistic characterisation of specific domains, adjectives and other parts of speech different from nouns are to be taken into consideration, especially in domains in which property concepts (i.e., adjectives) are a fundamental part of the discourse, like in tourism.

Numerous studies have been conducted in an effort to better understand the use, implications, and functions of adjectives and to classify them in clear-cut groups according to different criteria.[2] The main classification criteria employed so far are prototypical (Biber et al., 1999; Quirk et al., 1972), functional (Biber et al., 1999; Halliday, 1985), syntactic (Ferris, 1993; Teyssier, 1968), pragmatic (Kerbrat-Orecchioni, 1980), semantic (Dixon, 1982; Frawley, 1992; Hundsnurscher & Splett, 1982), or based on a combination of the above (Bertoldi et al., 2007; Fellbaum et al., 1993). Although different factors are considered to categorise this part of speech, there is a broad agreement to divide adjectives into two main groups, which are: descriptive (or objective) adjectives, whose purpose is to provide factual information and referential content, and evaluative (subjective) adjectives, whose goal is to evaluate and reflect a subjective viewpoint.[3] Descriptive and evaluative adjectives are related in that both display a person's perspective, which can be "physical (spatial and temporal), psychological, socio-cultural, and ideological" (Pierini, 2009, p. 98). Nonetheless, Pierini (2009) notes that descriptive adjectives can be further categorised since they provide factual (or objective) information about the referent, specifically a concrete property of the noun (e.g., *white*).

In the language of tourism in general, these predicative units, both descriptive and evaluative, play an important role and are fundamental to "present the beauty, allure, and uniqueness of destinations or of attractions" (Manca, 2016, p. 79). In Edo Marzá's (2011, p. 100) words, adjectives are often used with the aim of appealing to the reader, since they can express the view of the writer or the speaker and cause the reader to have some kind of feelings. Some previous research reveals meaningful results concerning the use of adjectives in this specialised domain. For example, Kang and Yu (2011) indicate that tourism texts in English use more adjectives than general texts, frequently implying a

---

2   For a comprehensive analysis of adjective classifications, see Heyvaert (2010) and Edo Marzá (2012, pp. 100–103).
3   Another classification of adjectives that is also frequent is that which divides them into two other groups: descriptive and relational adjectives, being the latter described as an adjective that does not express a property, but rather a relation to a concept designated by a noun (ten Hacken, 2019, p. 77), as for example in *architectural monument*.

positive meaning. In line with this finding, some studies like those carried out by Ding (2008), Edo Marzá (2011, 2012), and Salim et al. (2014) highlight a strong presence of positive and evaluative adjectives to express subjective judgements, to display positive emotions, and to portray outstanding qualities of accommodations, destinations, and so on. That is, adjectives significantly contribute to promotional communication in tourism and to the creation of destination image (Durán-Muñoz, 2019, p. 354; Pierini, 2009, p. 116) and, therefore, to the persuasive function of the language of tourism.

There has been some contrastive research published so far about the use of adjectives in this specialised domain combining different languages, such as English-Italian (Manca, 2008), English-Russian (Jaworska, 2013), and English-Spanish (Edo Marzá, 2012). However, most of the investigation conducted focuses on one single language and, hence, there is still an existing gap in the characterisation of this phenomenon from a contrastive viewpoint. To contribute to filling in this gap, the current research carries out a contrastive analysis with English and Spanish languages and aims to detect possible differences (and similarities) of both descriptive and evaluative adjectives in both languages when referring to activities, places, or people. These units are employed to profile products and services with the objective of attracting potential tourists and embellishing possible destinations. Nevertheless, in this specific subdomain, adjectives mainly highlight challenging and risky activities, besides nature and active involvement of participants, which differs from other subdomains (cf. Edo Marzá, 2011, 2012; Pierini, 2009) or tourism in general (Goethals & Segers, 2016; Jaworska, 2013). Examples (1) and (2)[4] show some evaluative adjectives in adventure tourism in both languages:

(1) Kick off your day with one of the most *exciting*, *adrenaline-rushing*, and *thrilling* activities we can think of.

(2) Ven y disfruta del rafting en Cantabria, donde un cañón *inolvidable* te hará disfrutar de este deporte de aventura, donde la diversión está asegurada.

As mentioned above, both descriptive and evaluative adjectives are profusely employed in this kind of promotional texts to describe natural spots, adventure activities, feelings, among others. However, the scope of this research is delimited to the contrastive study of descriptive adjectives in Spanish and in English. Particularly, the analysis focuses on the semantics of these adjectives, that is, the different semantic categories they evoke. In this line, Section 2.1. provides

---

4 All the examples of the chapter are taken from the ADVENCOR corpus. Italics are added by the authors.

some insights into semantic categorisations of adjectives, in general, and in the tourism domain, in particular.

2.1. Semantic categorisation of adjectives

The semantic function of adjectives refers to the qualities of things or, to put it differently, to "the attributes or properties associated with things" (Tucker, 1998, p. 57). Given the wide variety of these elements, different authors have proposed semantic classifications of adjectives regarding their features and behaviour (Peters & Peters, 2000).

In the category of descriptive adjectives, some distinctive dichotomies can be found (Heyvaert, 2010; Peters & Peters, 2000). Adjectives may be absolute or gradable, depending on whether they can be ordered according to a specific gradient feature (e.g., *tall*) or not (e.g., *dead*[5]); they may have a scalar meaning (like adjectives denoting a size, e.g., *big* or *small*); they may be classified as dynamic or stative in meaning if they refer to temporary conditions (e.g., *rude*) or not (e.g., *natural*), respectively; or they can be inherent (as in *heavy rock*) or non-inherent (as in *heavy drinker*). This last classification refers to the use of adjectives in context, and it depends on the elements of the sentence that are characterised. In the first case, the adjective indicates a property of the object itself (the *rock* is *heavy*), whereas in the second case, it is not the person but rather the person's activity that is described (the person's drinking is *heavy*) (Heyvaert, 2010, p. 1313).

Dixon (1982) was one of the first scholars who proposed a more elaborate semantic categorisation of adjectives. According to Raskin and Nirenburg (1995), the author suggests "a list of indispensable, must-have adjectives that even almost adjective-free languages, such as Chinese, Hausa, or Chinook, must somehow provide" (p. 7). This taxonomy of adjectives includes the following seven categories (Dixon, 1982, p. 16):

1. DIMENSION: *big, large, little, small; long, short; wide, narrow; thick, fat, thin,* and just a few more items.
2. PHYSICAL PROPERTY: *hard, soft; heavy, light; rough, smooth; hot, cold; sweet, sour,* to mention but a few.
3. COLOUR: *black, white, red,* and so on.

---

5 This type of adjectives may be used as gradable to convey a sense of irony (Kennedy, 1999).

4. HUMAN PROPENSITY: *jealous, happy, kind, clever, generous, gay, cruel, rude, proud, wicked*, among others.
5. AGE: *new, young, old*.
6. VALUE: *good, bad*, and a few more items (including *proper, perfect*, and perhaps *pure*, in addition to hyponyms of *good* and *bad*, such as *excellent, fine, delicious, atrocious, poor*, etc.).
7. SPEED: *fast, quick, slow*, and just a few more items.

In the language of tourism, we also find some categorisations of descriptive and, particularly, evaluative adjectives. These categories are mainly created ad hoc to cover the different types of adjectives that are found in a study and to fulfil its specific needs. For example, Pierini (2009) classifies the adjectives of her corpus-based study about accommodation into 16 semantic categories: "Aesthetic", "Appreciation", "Authenticity", "Availability", "Emotional impact", "Exclusiveness", "Internationality", "Money saving", "Newness", "Popularity", "Quantity", "Size", "Space", "Time", "Tradition", and "Wellness", and includes both evaluative and descriptive adjectives. In addition to this multi-class classification, others suggesting a lower number of main categories can be mentioned. For instance, the semantic categorisation proposed by Goethals and Segers (2016), which, in turn, is based on Edo Marzá's (2011, 2012) proposals, is divided into three different groups: evaluative, descriptive, and evaluative-descriptive. These categories are subsequently subdivided into some more. For example, the descriptive group includes geographical location, temporal location, description of physical and/or visible characteristics, description of intangible characteristics, and description of other characteristics that do not seem to imply an interpretative judgement on the part of the speaker and serve to specify the referential content. Finally, Pavlíčková and Rázusová (n.d.) follow the pragmatic classification of Kerbrat-Orecchioni (1980, as cited in Pavlíčková & Rázusová, n.d., pp. 7–8) (i.e., emotional, evaluative-non-axiological, and evaluative-axiological) to express the qualities by which British tour agencies and public institutions describe the country and the most remarkable sites to the public, as well as the way Slovak institutions do the same for English-speaking tourists. In this case, the adjectives analysed are all evaluative.

In this chapter, some categories of the three aforementioned proposals, that is, Dixon's (1982), Pierini's (2009), and Goethals and Segers's (2016) classifications, were selected (directly or slightly modified) to classify the descriptive adjectives selected for this research. Moreover, as Section 3.2. shows, further ad hoc categories were needed to complete the classification and to develop a suitable categorisation that attended to the specificities of this study. However, as Swales and Burke (2003) recognise, despite our efforts, both creating the categories and placing the adjectives in each of them are somewhat subjective processes in which certain reservations can always appear.

3. Methodology

The methodology employed in this research is assisted by the ADVENCOR corpus, a bilingual comparable corpus about adventure tourism (Durán-Muñoz & Jiménez-Navarro, 2021). Both subcorpora (English and Spanish) consist of promotional texts related to adventure tourism, published by public or private institutions or companies dedicated to tourism and originally written in the working languages, that is, translations were not considered. Specifically, the English subcorpus (1,064,664 words) is composed of texts in English from the United Kingdom, Ireland, and the United States, and the Spanish subcorpus (1,118,903 words) contains texts originally written in Spanish from Spain. The corpora were compiled semi-automatically using Sketch Engine[6] (Kilgarriff et al., 2004), although the entire compilation process was manually supervised to avoid irrelevant or inappropriate web pages and data to guarantee the quality of the results.

The corpus-assisted methodology is divided into three steps: (1) the extraction of candidate adjectives, (2) the selection and categorisation of these adjectives into two different groups: evaluative and descriptive, and (3) the classification of the descriptive adjectives subgroup according to their semantic meaning. The same process was undertaken for both languages, that is, English and Spanish, and it is depicted in Sections 3.1. and 3.2.

3.1. Extraction and selection of adjectives

The automatic extraction of the candidate adjectives from the corpus was carried out with the *Keywords* function of Sketch Engine[7] to calculate their keyness score. This option employs a hybrid method which combines statistical plus linguistic

---

6   https://www.sketchengine.eu/ [Last accessed: 28/09/2023].
7   To learn more about this function, please consult Kilgarriff (2009) and the following link: https://www.sketchengine.eu/my_keywords/keyword/#:~:text=Keywords%20are%20words%20(single%2Dtoken,reference%20corpus)%20or%20its%20subcorpus [Last accessed: 28/09/2023].

information to obtain successful results. A minimum frequency threshold of the candidates was set at five to limit the analysis to more frequent occurring items.

A list of 580 candidate adjectives for the English subcorpus and a list of 570 candidate adjectives for the Spanish subcorpus were produced automatically. Nevertheless, manual work was needed at this point to filter out false positives and to select the correct adjectival forms. Some candidates were discarded due to the following reasons:

1. They belonged to other part of speech categories (e.g., *itinerary*, *pedal* in English; *vertiente*, *río*, *acepto* in Spanish).
2. They were suffixes, prefixes, or were wrongly written (e.g., *multi-* and *re-* in English; *anti-*, *mini-*, and *audiovisual* in Spanish).
3. Derivative forms were wrongly lemmatised (e.g., *ferrata* and *ferratas* in English or *aventurero* and *aventurera* in Spanish). There are two interesting cases in this regard: on the one hand, those cases in which the plural form of the adjectives was included in the list but not the singular forms, like *difíciles* or *útiles* in Spanish; on the other hand, some noun forms were wrongly lemmatised to create adjectives that do not exist, like *gafo* from *gafas* also in Spanish.
4. They were part of proper names, like *Rican* (in *Costa Rican*) in English or *Cantábrico* (*Mar Cantábrico*) in Spanish.
5. They were adjective pronouns and possessives, like *your* in English or *nuestro* in Spanish.
6. They were extracted twice, such as *colour* and *color* (where the first one is British and the second one is American English) or *expres* and *exprés* in Spanish (where *expres* is wrongly written).

Moreover, since this study is limited to adjectives, nouns with a semantic adjectival status were discarded from the list (like *itinerary* in *itinerary details*). However, participles with a semantic adjectival status, such as *guided*, were included following Quirk and Greenbaum (1979) and Bosque Muñoz (1990), who state that the distinction between adjectives and participles is not always clear and depends on the verbal force of the latter.

As a result, the final list of adjectives selected for the analysis amounted to 510 for English and 406 for Spanish. Table 1 includes the top-10 units, ordered according to their keyness, to illustrate the adjectives selected in both languages:

Table 1. Top-10 adjectives in English and Spanish extracted from the ADVENCOR corpus with Sketch Engine ordered according to their keyness

Adjectives in EN	Frequency	Keyness	Adjectives in ES	Frequency	Keyness
tandem-j[8]	291	116.008	acuático-j	691	45.457
adventurous-j	210	36.979	aclaratorio-j	76	27.436
thrilling-j	184	34.493	aventurero-j	116	25.692
outdoor-j	1,330	33.073	náutico-j	264	25.467
alpine-j	137	26.706	hinchable-j	71	21.505
scenic-j	298	26.643	tibetano-j	69	21.32
unforgettable-j	140	22.236	estanco-j	61	20.09
multi-day-j	35	21.98	autoguiado-j	29	19.812
first-time-j	109	21.799	subacuático-j	55	18.98
downhill-j	62	21.354	rocoso-j	136	18.767

Once the adjectives were selected, the next step was their classification into the corresponding groups and subgroups according to their semantic meaning, as explained in Section 3.2.

3.2. Classification of the selected adjectives

The selected adjectives were first classified under the following two categories:

1. Evaluative adjectives, which are a means for subjective evaluation that are often used by the writer or speaker to persuade potential visitors and to provide what qualities make the object of interest worth visiting. Adjectives included in descriptions which depend on the speaker's/writer's viewpoint, such as in *a short hike*, are also considered evaluative.
2. Descriptive adjectives, which are concerned with objective features, such as colours, locations, geographical features, and so forth, and provide tourists (or potential tourists) with descriptions full of details to help understand what they will see during their trips.

---

8 The "-j" element is added by the programme to indicate that the corresponding units are adjectives.

After the adjectives were classified into these two groups, the semantic meaning of the descriptive adjectives was analysed in the two working languages, and they were categorised according to 14 semantic categories and their corresponding subcategories, which are shown in Table 2 along with a brief description.[9]

Table 2. Proposal of categorisation for descriptive adjectives extracted from the ADVENCOR corpus

SEMANTIC CATEGORY	BRIEF DESCRIPTION
Abstract/Intangible characteristics	Attributes that cannot be physically touched or seen but hold significance or value in certain contexts (e.g., *certified*)
Accommodation	Units related to the places where travellers can stay during their vacations (e.g., *en-suite*)
Adventure activity-related	Terms associated with exciting or thrilling activities that involve exploration, physical challenges, or unique experiences (e.g., *climbable*)
Area-related	Terms describing an area or a place (e.g., *uninhabited*)
Fauna	Terms that refer to animals (e.g., *amphibious*)
Human-related	Attributes related to inherent inclination or tendency of individuals towards certain behaviours, preferences, or characteristics (e.g., *vegetarian*)
Level of experience	Degree of knowledge, skill, and familiarity that an individual needs for a particular activity (e.g., *technical*)
Money-related	Terms associated with financial aspects or transactions (e.g., *refundable*)
Origin	Units that refer to places where something begins or originates (e.g., *Basque*)

(*continued*)

---

[9] The tables with the classification of the descriptive adjectives in English and Spanish are provided in Appendixes 1 and 2, respectively.

Table 2. Continued

SEMANTIC CATEGORY	BRIEF DESCRIPTION
Physical, visible characteristics: Body-related Climate-related Colour Landscape description Manufacture Material Motion-related Physical property Protection-related Shape	Qualities or attributes that are evident through visual perception or touch (e.g., *colourful*)
Position/Geographical location	Attributes related to placement or a specific area in which something is situated or occurs (e.g., *indoor*)
Quantification	Terms that express numerical relationships, proportions, or divisions (e.g., *half*)
Temporal location	Units referring to a time period or era in which something exists or occurs (e.g., *prehistoric*)
Temporality	Attributes related to the time or the duration of events, actions, or objects (e.g., *one-day*)

As mentioned in Section 2.1., the resulting list of categories and subcategories was a combination of some categories taken (directly or modified) from three proposals, that is, Dixon's (1982), Pierini's (2009), and Goethals and Segers's (2016), plus several ad hoc categories that were needed to cover all the descriptive units selected for this specific domain.

As an illustration, the categories "Abstract/Intangible characteristics", "Money-related", "Physical, visible characteristics", and "Temporal location" were retrieved from Goethals and Segers's (2016), while the "Human propensity" (in this study, "Human-related") category and the "Colour" and "Physical property" subcategories were borrowed from Dixon's (1982). There were also some adjustments in other categories, such as the original "Geographical location" category from Goethals and Segers's (2016) proposal, which was extended by adding the "Position" feature as considered to be needed and decidedly related. Moreover, the "Authenticity", "Quantity", and "Time" categories were found in Pierini's (2009) study about adjectives in the tourism sector, which was supportive, but renamed "Origin", "Quantification", and "Temporality" for semantic reasons related to the arranged adjectives.

Finally, taking into consideration the specific semantics and nature-related tendency of the language of adventure tourism, the categories and subcategories created ad hoc were: (1) categories: "Accommodation", "Adventure activity-related", "Area-related", "Fauna", and "Level of Experience"; and (2) subcategories: "Body-related", "Climate-related", "Landscape description", "Manufacture", "Material", "Motion-related", "Protection-related", and "Shape".

At this point, it is worth noting what some scholars (Marx, 1977, 1983; Szalay & Deese, 1978, both cited in Raskin & Nirenburg, 1995, p. 21) refer to the "plasticity" of adjectival meaning, that is, the same adjective can emphasise a different property of a noun in a different context. For example, the adjective *green* has different meanings in the expressions *greenhouse* and *green T-shirt*, since nouns influence the meaning of the adjective. Therefore, this was taken into account during the semantic classification of the adjectives and their meaning in context was checked when there were uncertainties, as illustrated in Section 4.

4. Analysis and Results

The most relevant results of this English-Spanish contrastive study are described in this section. Firstly, we depict the classification of adjectives as evaluative and descriptive; then, we focus on the classification and features of the descriptive adjectives, paying particular attention to differences and similarities between the two working languages.

4.1. Analysis of evaluative and descriptive adjectives

In the first place, the selected adjectives were divided into two large groups: descriptive and evaluative. At first glance, this process appears elementary, since there is a notable difference between these two groups: descriptive adjectives refer to objective facts, while evaluative adjectives imply subjective assessment to some extent. Most adjectives were classified at once due to their clarity (e.g., *triple* [descriptive] or *beautiful* [evaluative]).

Nonetheless, this classification resulted in a rather arduous and lengthy process because of several adjectives which gave rise to different interpretations and confusion. This was the case, for example, of the adjective *recóndito* in Spanish and its equivalent in English, *secluded*, both of which were finally considered to be descriptive. In order to clarify cases like these, contexts were checked using the *Concordance* function in Sketch Engine, which was helpful in numerous instances to differentiate descriptive and evaluative meanings. Even then, there were various adjectives which had to be double-checked with the aim of reaching a consensus. This is the case of *beaten* and *level* in English. To do so, contexts like (5) and (6) were examined:

(5) Mountain Biking and Hiking in Arkansas Adventure starts off the *beaten* path.
(6) Cottage plot is *level*, grassy and sunny.

Context (5) clearly shows that the adjective *beaten* is employed here with an evaluative meaning, indicating it is not well travelled. On the other hand, the context of *level* indicates that this term is employed here as an adjective and that it has a descriptive meaning related to position.

Other cases that also required some double check in the corpus via the *Concordance* function were those with several descriptive meanings, like the adjective *rustic*. This was classified into two different groups ("Accommodation" and "Landscape description") due to its use in context, as shown in examples (7) and (8):

(7) From *rustic* cabins we'll explore Volcanoes National Park through forests of ferns, orchids, wild fruits, and rare birds.
(8) You'll temporarily wonder if we accidentally teleported to Europe, as the deep pine forests and *rustic* cow pastures are a real novelty!

In Spanish there were also some similar cases. As a way of example, we found *serrano*, which can have several meanings. In this case, all the meanings of this adjective are descriptive, but they could be classified in one or another category depending on its semantic meaning. Nevertheless, all the cases refer to a mountain range, as in the contexts (9) and (10).

(9) En esta ruta circular de montaña podremos ascender al macizo del Mondalindo y disfrutar de la belleza de las panorámicas del entorno *serrano*.
(10) Turismo ornitológico, micológico, arquitectónico, rutas por los pueblos más hermosos de nuestra sierra, descubriendo la tradición y la cultura *serrana*.

The final list of descriptive and evaluative adjectives in the two working languages, English and Spanish, confirms that the presence of both types of adjectives is high in both subcorpora. On the one hand, in English, 510 adjectives were obtained, of which 290 were evaluative (56.9 %) and 220 descriptive (43.1 %). On the other hand, a total of 406 adjectives were selected in Spanish, 267 descriptive (65.7 %), and 139 evaluative (34.3 %). Of interest here is the difference in number between English and Spanish despite the size of both subcorpora. In other words, even though the Spanish subcorpus is larger than the English subcorpus by more than 100,000 words, the number of adjectives detected in the English one is much higher (510 vs 406, respectively). This reveals an unanticipated result about the use of adjectives in both languages. It was initially expected that the use of adjectives in both languages in this specialised domain would be approximately the same. Nevertheless, this finding highlights other two important facts: first, the English language is more prolific in the use of adjectives in general than Spanish, which employs a lesser number of adjectives; and second, evaluative adjectives are also more frequent in English than in Spanish in this subdomain, which denotes a higher preference for a more persuasive style. In fact, this tendency in both languages is observed in Table 1, which contains the top-10 adjectives from the corpus ordered according to their keyness. In English, we observe four evaluative adjectives out of 10 (*adventurous, thrilling, scenic,* and *unforgettable*), while in Spanish there are only two (*aclaratorio* and *aventurero*).

In turn, when the adjectives are ordered according to their frequency (cf. Table 3), the number of evaluative adjectives in both languages differs. In English, the number of evaluative adjectives grows greatly compared to Table 1 and we find seven different terms with this type of meaning (*high, beautiful, long, full, easy, amazing,* and *short*), while in Spanish three different evaluative adjectives are detected: *alto, necesario,* and *diferente*. In both languages the number of evaluative adjectives when they are ordered according to their frequency is greater than when they are ordered following their keyness, especially in English; however, the number in the English language is greater again. This is also a relevant result for other two main reasons: first, it states that those adjectives with a greater keyness are mostly descriptive, especially in Spanish; second, it proves that the keyness of the adjectives extracted from the corpus is not correlated with their frequency in most cases, since only one adjective is repeated for each language (*outdoor* in English and *acuático* in Spanish) if we compare Tables 1 and 3.

Table 3. Top-10 adjectives in English and Spanish extracted from the ADVENCOR corpus with Sketch Engine ordered according to their frequency

Adjectives in EN	Frequency	Keyness	Adjectives in ES	Frequency	Keyness
high-j	1,346	2.005	*natural*-j	1,969	7.032
outdoor-j	1,330	33.073	*activo*-j	1,532	13.408
available-j	915	2.194	*deportivo*-j	1,127	7.494
beautiful-j	835	6.55	*alto*-j	1,104	1.944
long-j	716	2.023	*necesario*-j	904	2.155
full-j	672	2.081	*libre*-j	761	3.018
free-j	652	2.091	*rural*-j	738	7.705
easy-j	636	2.463	*medio*-j	725	1.889
amazing-j	549	7.551	*acuático*-j	691	45.457
short-j	491	2.311	*diferente*-j	648	2.086

For a closer look at the results, the first 50 adjectives in both languages (ordered according to their keyness) are analysed and compared, and surprising results are also obtained throughout. Firstly, both descriptive and evaluative adjectives are found almost equally represented in the first 50 adjectives in the English list, although the descriptive ones are slightly more numerous (54 %). This means that the usage of evaluative adjectives in this language is remarkably frequent, both to indicate positive aspects, like *beautiful, amazing, perfect, unique, exciting, excellent*, or *spectacular*, and to provide evaluative descriptions, such as *high, long, full, easy*, or *short* (cf. example (9)).[10]

(9) The leader and group should have *easy* access to emergency and contingency equipment.

The list of Spanish adjectives shows a completely different result. They both coincide in the fact that descriptive adjectives outnumber evaluative adjectives, but in the Spanish list only seven evaluative adjectives are found among the top-50. Therefore, contrary to the English list, descriptive adjectives represent the 86 % of the total adjectives, which means that evaluative adjectives are not that common when they are ordered according to their keyness.

However, when the top-50 adjectives are ordered in accordance with their frequency, the results are exactly the opposite. In both languages, 64 % of the adjectives are evaluative, while only 36 % are descriptive. This reveals another meaningful finding of the study, which refers to the fact that descriptive adjectives are more specialised in this domain than evaluative adjectives and, thus, they convey the special meaning.

Finally, the total number of descriptive adjectives in both subcorpora is slightly superior to the number of evaluative adjectives, as shown in Table 4. This fact responds to the first research question of this work, that is, which type of adjective (descriptive or evaluative) is more common in each language under analysis. Moreover, it leads to our second and third research questions, that is, what semantic categories are more frequent among the descriptive adjectives and what are the main differences in their usage in the two languages. These questions are addressed in Section 4.2.

---

10 We think *evaluative descriptions* are a kind of description which depends on the speaker's point of view and, therefore, are subjective. In this case, adjectives that are part of an evaluative description are also considered evaluative.

Table 4. Total frequency of both descriptive and evaluative adjectives in both languages

	Descriptive	Evaluative
English	220	290
Spanish	267	139
Total	487; 53 %	429; 47 %

4.2. Categorisation of descriptive adjectives in English and Spanish

The categorisation of descriptive adjectives in both languages is included in Appendix 1 (English) and Appendix 2 (Spanish). They show that all the categories proposed in the study are filled in with adjectives in both languages.

The adjectives were distributed among the categories according to not only their meaning but also their usage in context. As with the classification of descriptive and evaluative adjectives described in Section 4.1., the semantic categorisation was straightforward in most of the cases; however, some adjectives were confusing, and a further analysis of their contexts was required. This is the case of *glacial* in English and *glaciar* in Spanish, which could be classified under "Landscape description" or "Climate-related" categories. Curiously, all the cases found for these adjectives were employed in the same way in English and in Spanish, and all of them referred to "Landscape description". In English, the unit was combined with terms like *river, valley, terrain, mountain*, and so on, as in example (10), and in Spanish this combination was made with terms such as *frente, valle, terreno*, among others, as in example (11).

(10) Explore the South Island's legendary peaks, *glacial* lakes and rugged coastline on foot, by boat and by bike.

(11) Nos situaremos justo encima del frente *glaciar* para poder disfrutar de la visión del hielo cayendo al mar.

The category with more adjectives in English was "Position/Geographical location" with 33 types (vs 23 in Spanish). However, in Spanish was "Origin", with 45 types (vs 25 in English). The distribution of adjectives also unveiled some features of both languages in this tourism subdomain, as shown in Figure 1.

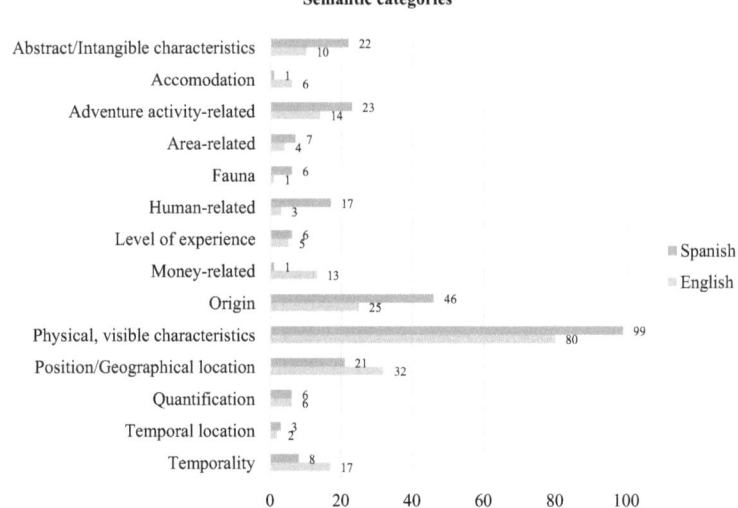

Figure 1. Distribution of descriptive adjectives in Spanish and English

Firstly, most of the descriptive adjectives in Spanish were gathered in four main categories, that is, "Physical, visible characteristics", with 99 types; "Origin", with 46 types; "Adventure activity-related", with 23; and "Abstract/Intangible characteristics", with 22. In turn, only two categories in English presented more than 30 adjectives, namely, "Physical, visible characteristics", with 80 types, and "Position/Geographical location", with 32. Within the rest of the English categories the distribution was more balanced. Secondly, it is noteworthy that the category "Money-related" has 13 adjectives in English but only one in Spanish. The category "Human-related" also attracts our attention, since only three occurrences were found in English but 17 in Spanish, which can lead to the conclusion that, in Spanish, promotional texts in this subdomain address people (*organizador*, *acompañante*, *aventurero*, etc.) more than in English, where they can be rather impersonal and general. Finally, another interesting result is about the "Colour" subcategory (belonging to the "Physical, visible characteristics" category), whose adjectives are double in English than in Spanish, and also the type of colours that are used are more varied and not that frequent, like *verdant*, *azure*, *carroty*, or *aqua* in English, compared to *turquesa* or *rodeno* in Spanish. This is again a feature of the language of adventure tourism, used with the aim of incorporating some poetic elements and, therefore, embellishing it.

Regarding specificities of the languages, English and Spanish share some lexical similarities. For example, many direct equivalents are found in the different categories, both sharing morphological similarity, such as *elastic* (in English) and *elástico* (in Spanish), *adjustable* (in English) and *ajustable* (in Spanish), *mountainous* (in English) and *montañoso* (in Spanish), and not sharing it, like *sandy* (in English) and *arenoso* (in Spanish), *level* (in English) and *llano* (in Spanish), *waterproof* (in English) and *impermeable* (in Spanish). To put it differently, there is a great number of adjectives that find their equivalents in the other subcorpus and, consequently, we can state that adjectives employed to describe natural environments or adventure activities are substantially similar in both languages.

Finally, with respect to Spanish adjectives, the use of foreign words, like *outdoor* and *indoor*, should be highlighted. This supports the idea that the use of anglicisms in other languages, like Spanish, is another technique employed in the language of adventure tourism (and the language of tourism in general) to incorporate some attractive, exotic, and persuasive elements in this specialised discourse (Durán-Muñoz, 2019).

5. Conclusions

This corpus-assisted contrastive research on adjectivisation contributes to the characterisation of the language of adventure tourism by analysing an understudied but fundamental aspect of it, that is, adjectives. By doing so, it provides answers to the initial questions of this research.

Regarding the first question (which type of adjective, that is, descriptive or evaluative, is more frequent in this subdomain), we can confirm the important role played by both descriptive and evaluative adjectives when referring to activities, places, or people in the two working languages. Initially, it was expected that the use of adjectives in both languages would be approximately the same. However, English and Spanish proved to work differently with respect to adjectives. Despite the fact that the Spanish subcorpus is over one hundred thousand words larger than the English subcorpus, the number of detected adjectives is higher in the English subcorpus. This indicates that the English language employs adjectives more frequently than Spanish, making it more prolific in this respect. Additionally, it employs evaluative adjectives with a higher frequency too, which suggests a greater inclination towards employing persuasive language through the use of adjectives.

With respect to the second question (what semantic categories are more frequent among the descriptive adjectives in the corpus), adjectives in this subdomain were classified into 14 categories according to their semantic meaning. We could see that all the categories outlined in the study contain adjectives in both languages. In terms of frequency, the "Origin" category was the most prominent in Spanish, whereas the "Position/Geographical location" category was the densest in English. The distribution of adjectives in both languages also revealed some noteworthy differences in this tourism subdomain, as mentioned in Section 4.2., which answers our third research question, that is, what are the main differences in their usage in the languages under study.

Finally, in some instances, the context in which adjectives were used proved to be essential for distinguishing their intended meaning. It was necessary to determine whether these adjectives were evaluative or descriptive, and then to accurately classify them into their respective semantic category. The information gleaned from observing the adjectives in their real context, alongside their co-text and any collocations, proved to be highly valuable.

To conclude, adjectivisation is a truly essential part of the language of adventure tourism, and it is worth studying it by means of either a monolingual or a contrastive study like this one. Representative corpora like ADVENCOR provide interesting and (sometimes) surprising information that help understand specialised languages, build specialised resources, such as *DicoAdventure* (Durán-Muñoz, 2021),[11] and improve the quality of translation products. As to future lines of research, there is still much to do in this field, like the study of evaluative adjectives and their semantic categorisation, a comparative analysis of descriptive and/or evaluative adjectives in this subdomain with other subdomains, and the study of specialised collocations with adjectives as their base.

References

Alonso, A., & Torner, S. (2010). Adjectives and collocations in specialized texts: Lexicographical implications. In A. Dykstra & T. Schoonheim (Eds.), *Proceedings of the 14th EURALEX International Congress* (pp. 872–881). Fryske Akademy – Afûk.

ATTA (Adventure Travel Trade Association). (2022). *Adventure travel guide qualifications & performance standard – III. definitions*. https://www.adventuretravel.biz/education/adventure-edu/guide-standards/adventure-travel-guide-qualifications-performance-standard-iii-definitions/#:~:text=Adventure%20travel%20is%20a%20type,specialized%20skills%20and%20physical%20exertion

---

11 http://olst.ling.umontreal.ca/dicoadventure/ [Last accessed: 28/09/2023].

Bertoldi, A., De Oliveira Chishman, R. L., & Da Rosa Alves, I. M. (2007). Adjectival semantics and the legal domain: A study for ontology improvement. *Inteligencia Artificial*, *11*(36), 19–26. https://doi.org/10.4114/ia.v11i36.887

Biber, D., Johansson, S., Leech, G., Conrad, S., & Finegan, E. (1999). *Longman grammar of spoken and written English*. Longman.

Bosque Muñoz, I. (1990). *Las categorías gramaticales*. Editorial Síntesis SA.

Buckley, R. (2006). *Adventure tourism*. Digital Library CABI.

Buckley, R. (2010). *Adventure tourism management*. Butterworth-Heinemann.

Cappelli, G. (This volume). The language of accessible adventure tourism. In I. Durán-Muñoz & E. L. Jiménez-Navarro (Eds.), *Exploring the language of adventure tourism: A corpus-assisted approach*. Peter Lang.

Ding, D. G. (2008). *Features and tourism English and its translation*. Shanghai Jiao Tong University Press.

Dancette, J., & L'Homme, M. C. (2004). Building specialized dictionaries using lexical functions. *Linguistica Antverpiensia, New Series – Themes in Translation Studies*, *3*, 113–131. https://doi.org/10.52034/lanstts.v3i.107

Dixon, R. M. W. (1982). *Where have all the adjectives gone? And other essays in semantics and syntax*. Mouton. https://doi.org/10.1515/9783110822939

Durán-Muñoz, I. (2014). Aspectos pragmático-lingüísticos del discurso del turismo de aventura: Estudio de un caso. *Normas: revista de estudios lingüísticos hispánicos*, *4*, 49–69. https://doi.org/10.7203/Normas.4.4687

Durán-Muñoz, I. (2019). Adjectives and their keyness. A corpus-based analysis in English tourism. *Corpora*, *14*(3), 1749–5032. https://doi.org/10.3366/cor.2019.0178

Durán-Muñoz, I. (2021). *DicoAdventure* y la terminología del turismo de aventura: Propuesta de diccionario en línea. In T. Barceló Martínez, I. Delgado Pugés, & F. García Luque (Eds.), *Tendencias actuales en traducción especializada, traducción audiovisual y accesibilidad* (pp. 395–417). Tirant Lo Blanch.

Durán-Muñoz, I., & L'Homme, M. C. (2020). Diving into adventure tourism from a lexico-semantic approach: An analysis of English motion verbs. *Terminology*, *26*(1), 33–59. https://doi.org/10.1075/term.00041.dur

Durán-Muñoz, I., & Jiménez-Navarro, E. L. (2021). Colocaciones verbales en el turismo de aventura: Estudio contrastivo inglés-español. In G. Corpas Pastor, M.ª R. Bautista Zambrana, & C. M. Hidalgo-Ternero (Eds.), *Sistemas fraseológicos en contraste: Enfoques computacionales y de corpus* (pp. 121–142). Comares.

Edo Marzá, N. (2011). A comprehensive corpus-based study of the use of evaluative adjectives in promotional hotel websites. *Odisea, 12*, 97–123. https://doi.org/10.25115/odisea.v0i12.222

Edo Marzá, N. (2012). Páginas web privadas e institucionales: El uso de la adjetivación en un corpus inglés-español de promoción de destinos turísticos. In J. Sanmartín Sáez (Ed.), *Discurso turístico e Internet* (pp. 81–124). Iberoamericana/Vervuert.

Fellbaum, C., Gross, D., & Miller, K. (1993). Adjectives in WordNet. In G. A. Miller (Ed.), *Five papers on WordNet* (pp. 26–39). Princeton University. https://wordnetcode.princeton.edu/5papers.pdf

Ferris, C. D. (1993). *The meaning of syntax. A study in the adjectives of English.* Longman.

Frawley, W. (1992). *Linguistic semantics.* Lawrence Erlbaum Associates.

Goethals, P., & Segers, L. (2016). El uso de los adjetivos en los folletos de Turespaña y en la guía de viajes 2.0 Minube: Un análisis de corpus. In M. López Santiago & D. Giménez Folqués (Eds.), *El léxico del discurso turístico 2.0* (pp. 117–152). Universitat de València.

Halliday, M. A. K. (1985). *An introduction to functional grammar.* Arnold.

Heyvaert, F. (2010). An outline for a semantic categorization of adjectives. In A. Dykstra & T. Schoonheim (Eds.), *Proceedings of the 14th EURALEX International Congress.* Fryske Akademy – Afûk.

Hudson, S. (2003). *Sport and adventure tourism.* The Haworth Press.

Hundsnurscher, F., & Splett, J. (1982). *Semantik der Adjektive im Deutschen: Analyse der semantischen Relationen.* Verlag.

International Organization for Standarization (ISO). (2009). *ISO 704: 2009. Terminology work. Principles and methods.* ISO.

Janowski, I., & Reichenberger, I. (2019). Conceptualising adventure tourism from a consumer perspective. *Travel and Tourism Research Association: Advancing Tourism Research Globally, 3*, 1–7. https://scholarworks.umass.edu/ttra/2019/grad_colloquium/3

Jaworska, S. (2013). The quest for the "local" and "authentic" corpus-based explorations into the discursive constructions of tourist destinations in British and German commercial travel advertising. In D. Höhmann (Ed.), *Tourismuskommunikation. Im Spannungsfeld von Sprach- und Kulturkontakt* (pp. 75–100). Peter Lang.

Kang, N., & Yu, Q. (2011). Corpus-based stylistic analysis of tourism English. *Journal of Language Teaching and Research, 2*(1), 129–136. https://doi.org/10.4304/jltr.2.1.129-136

Kennedy, C. (1999). *Projecting the adjective. The syntax and semantics of gradability and comparison*. Routledge.

Kerbrat-Orecchioni, C. (1980). *L'énonciation de la subjectivité dans le langage*. Armand Colin.

Kilgarriff, A. (2009). Simple maths for keywords. In M. Mahlberg, V. González Díaz, & C. Smith (Eds.), *Proceedings of the Corpus Linguistics Conference (CL2009)* (pp. 1–6). University of Liverpool.

Kilgarriff, A., Rychlý, P., Smrž, P., & Tugwell, D. (2004). The Sketch Engine. In G. Williams & S. Vessier (Eds.), *Proceedings of the 11th EURALEX International Congress, EURALEX 2004. Volume I* (pp. 105–116). UBS – Université de Bretagne-Sud.

Lee, T. H., Tseng, C. H., & Jan, F. H. (2015). Risk-taking attitude and behavior of adventure recreationists: A review. *Journal of Tourism and Hospitality*, 4(2), 1–3. https://doi.org/10.4172/2167-0269.1000149

Leech, G. (1989). *An A–Z of English. Grammar & usage*. Nelson.

L'Homme, M. C. (2002, August 28–30). *What can verbs and adjectives tell us about terms?* [Conference presentation]. 6th International Conference on Terminology and Knowledge Engineering Conference (TKE), Nancy, France.

Lorente, M. (2001). Terminología y fraseología especializada: Del léxico a la sintaxis. In M. Pérez & G. Guerrero (Eds.), *Panorama actual de la Terminología* (pp. 159–180). Comares.

Manca, E. (2008). From phraseology to culture. Qualifying adjectives in the language of tourism. *International Journal of Corpus Linguistics*, 13(3), 368–385. https://doi.org/10.1075/ijcl.13.3.07man

Manca, E. (2016). *Persuasion in tourism discourse: Methodologies and models*. Cambridge Scholars Publishing.

Marx, W. (1977). Die Kontextabhängigkeit der Assoziativen Bedeutung. *Zeitschrift für experimentelle und angewandte Psychologie*, 24, 455–462.

Marx, W. (1983). The meaning-confining function of the adjective. In G. Rickheit & M. Bock (Eds.), *Psycholinguistic studies in language processing* (pp. 70–81). Walter de Gruyter.

Muller, T. E., & Cleaver, M. (2000). Targeting the CANZUS baby boomer explorer and adventurer segments. *Journal of Vacation Marketing*, 6(2), 154–169. https://doi.org/10.1177/135676670000600206

Page, S. J., Bentley, T. A., & Walker, L. (2005). Scoping the nature and extent of adventure tourism operations in Scotland: How safe are they? *Tourism Management*, 26(3), 381–397. https://doi.org/10.1016/j.tourman.2003.11.018

Pavlíčková, E., & Rázusová, M. (n.d.). *Comparative analysis of evaluative adjectives in tourism texts*. Academia. https://www.academia.edu/3575244/Comparative_analysis_of_evaluative_adjectives_in_tourism_texts

Peters, I., & Peters, W. (2000). The treatment of adjectives in SIMPLE: Theoretical observations. In M. Gavrilidou et al. (Eds.), *Proceedings of the Second International Conference on Language Resources and Evaluation (LREC'00)* (pp. 1–8). European Language Resources Association (ELRA).

Pierini, P. (2009). Adjectives in tourism English on the web: A corpus-based study. *CÍRCULO de Lingüística Aplicada a la Comunicación*, *40*, 93–116. https://revistas.ucm.es/index.php/CLAC/article/download/41886/39904/58746

Pitkänen-Keikkilä, K. (2015). Adjectives as terms. *Terminology*, *21*(1), 76–101. https://DOI.org/10.1075/term.21.1.04pit

Quirk, R., & Greenbaum, S. (1979). *A university grammar of English*. Longman Group Limited.

Quirk, R., Greenbaum, S., Leech G., & Svartvik, J. (1972). *A Grammar of Contemporary English*. Longman Group Limited.

Raskin, V., & Nirenburg, S. (1995). Lexical semantics of adjectives. A microtheory of adjectival meaning. *MCCS-95-288. Memoranda in Computer and Cognitive Science* (pp. 1–69). Computing Research Laboratory. https://www.researchgate.net/publication/2314119_Lexical_Semantics_of_Adjectives_A_Microtheory_Of_Adjectival_Meaning

Salim, M., Ibrahim, N., & Hassan, H. (2014). Promoting diversity via linguistic and visual resources: An analysis of the Malaysian tourism website. *LSP International*, *1*, 1–14.

Sung, H., Morrison, A., & O'Leary, J. (2000). Segmenting the adventure travel market by activities: From the North American industry providers' perspective. *Journal of Travel & Tourism Marketing*, *9*(4), 1–20. https://doi.org/10.1300/J073v09n04_01

Swales, J. M., & Burke, A. (2003). 'It's really fascinating work': Differences in evaluative adjectives across academic registers. In P. Leistyna & C. F. Meyer (Eds.), *Corpus analysis: Language structure and language use* (pp. 1–18). Rodopi.

Swarbrooke, J., Beard, C., Leckie, S., & Pomfret, G. (2003). *Adventure tourism: The new frontier*. Routledge.

Szalay, L. B., & Deese, J. (1978). *Subjective meaning and culture: An assessment through word association*. Erlbaum.

ten Hacken, P. (2019). Relational adjectives between syntax and morphology. *SKASE Journal of Theoretical Linguistics*, *19*(1), 77–92.

Teyssier, J. (1968). Notes on the syntax of the adjective in Modern English. *Lingua*, *20*, 225–249.

Tucker, G. H. (1998). *The lexicogrammar of adjectives a systemic functional approach to lexis*. Cassell.

UNWTO. (2014). *World Tourism Organization: Annual Report 2013*. http://www2.unwto.org/publication/unwto-annual-report-2013

Acknowledgements

This work has been carried out within the framework of the R&D project "DicoAdventure: diseño y desarrollo de un recurso electrónico especializado bilingüe (inglés, español) sobre el turismo de aventura a partir de marcos semánticos" (Ref. UCO-1380857-F), co-funded by the Operational Programme FEDER 2014–2020 and the Consejería de Economía, Conocimiento, Empresas y Universidad of the Andalusian regional government. Partially, it has also been funded by VIP II (PID2020-112818GB-I00) and Recover (Ref. ProyExcel_00540).

Appendix 1. Semantic categorisation of English descriptive adjectives

Semantic Category	Adjectives	
Abstract/ Intangible characteristics	*breathable, certified, English-speaking, exposed, interpretive, licensed, living, optional, scheduled, spare*	
Accommodation	*en-suite, family-run, lodge-based, non-residential, pet-friendly, rustic*	
Adventure activity-related	*acrobatic, aerobatic, all-terrain, climbable, equestrian, equipped, geologic, navigable, nautical, self-guided, speleological, touristic, unmarked, unweighted*	
Area-related	*undiscovered, unexplored, uninhabited, populated*	
Fauna	*amphibious*	
Human-related	*handicapped, nomadic, vegetarian*	
Level of experience	*first-time, introductory, non-technical, Olympic, technical*	
Money-related	*add-on, affiliated, all-inclusive, discounted, full-service, insured, non-commercial, non-refundable, payable, rental, refundable, toll-free, uninsured*	
Origin	*Aboriginal, Alpine, Alaskan, Arctic, Atlantic, Austrian, Balinese, Basque, Burmese, Caribbean, Croatian, Dominican, Hawaiian, Mediterranean, Mongolian, Nepalese, Nepali, Nordic, Siberian, Slovenian, Swiss, Thai, Tibetan, Tyrolean, Welsh*	
Physical, visible characteristics	Body-related	
	barefoot, foot-launched	
	Climate-related	
	downwind, heated, polar, rainy, shaded, snow-covered, snow-capped, snowy, soarable, sub-tropical (subtropical), sunny, thermal, tropical, wet	
	Colour	
	aqua, azure, carroty, colorful (AmE) (colourful (BrE)), coral, emerald, iridescent, red-blazed, turquoise, verdant	
	Landscape description	
	arboreal, boreal, cascading, choppy, coastal, craggy, deciduous, forested, glacial, grassy, hilly, jagged, level, marine, mountainous, muddy, nature-related, panoramic, paved, rocky, rugged, rustic, sandy, secluded, sloping, steep, starry, volcanic, wooded	

Semantic Category	Adjectives
	Manufacture
	bottled, customized, homemade, man-made, tailored, tailor-made
	Material
	stainless, woollen
	Motion-related
	airborne, semi-static, slippery, static
	Physical property
	adjustable, collapsible, convertible, elastic, folding, non-motorized, rigid, stretchy, sturdy, waterproof, windproof
	Protection-related
	covered, sheltered
Position/ Geographical location	*aerial, aquatic, backward, cross-country, downhill, easterly, east, halfway, indoor, inland, locally-sourced, neighboring, north, northern, one-way, on-site, open-air, outdoor, southwestern, southeast, south, stand-up, subterranean, surrounding, tree-top, underground, underfloor, upstate, upstairs, uphill, underwater, vertical*
Quantification	*double, half, maximum, solo, triple, twin*
Temporal location	*prehistoric, Jurassic*
Temporality	*all-day, bi-annual (biannual), daylong (day-long), durable, four-day, last-minute, limitless, multi-day, one-day, one-time, overnight, round-trip, seasonal, three-day, two-day, two-hour, year-round*

Appendix 2. Semantic categorisation of Spanish descriptive adjectives

Semantic Category	Adjectives
Abstract/ Intangible characteristics	*activo, adepto, alfanumérico, económico-social, endémico, festivo, instructivo, interpretativo, isotónico, libre, media-alta, nominativo, opcional, obligatorio, sociocultural, subjetivo, temático, tutelar, turístico, turístico-deportivo, unisex, vacacional*
Accommodation	*supletorio*
Adventure activity-related	*acrobático, aerostático, autoguiado, botánico, cartográfico, cinegético, deportivo, didáctico, ecuestre, ecológico, espeleológico, etnográfico, físico, físico-deportivo, florística, hípico, náutico, recreativo, respirador, todoterreno, topográfico, velero*
Area-related	*accesible, cosmopolita, inaccesible, inexplorado, inhóspito, recóndito, virgen,*
Fauna	*faunístico, mamífero, nival, ornitológico, rapaz, zoológico*
Human-related	*acompañante, aventurero, barranquista, ciclista, colaborador, conocedor, discapacitado, explorador, instructor, montañero, nómada, organizador, paralítico, peregrino, pionero, senderista, soltero*
Level of experience	*especialista, experto, inexperto, novato, principiante, técnico*
Money-related	*promocional*
Origin	*aborigen, alpino, ártico, astur, asturiano, atlántico, autóctono, alicantino, aragonés, balear, canario, cántabro, cantábrico, catarinense, castellano, conquense, esquimal, finlandés, gallego, granadino, grancanario, ibérico, inca, islandés, levantino, leonés, mediterráneo, montano, montés, morisco, maya, nepalí, navarro, nórdico, noruego, oscense, palentino, pirenaico, provenzal, serrano, sherpa, somontano, tirolés, tibetano, valenciano, vikingo*

Descriptive adjectives in adventure tourism

Semantic Category	Adjectives	
Physical, visible characteristics	Body-related	
	facial, labial	
	Climate-related	
	alisio, bravo, climatológico, húmedo, medioambiental, meteorológico, nevado, nuboso, polar, seco, solar, termal, tropical, umbrío, ventoso	
	Colour	
	moreno, rodeno, rubio, turquesa, verde	
	Landscape description	
	abrupto, agreste, arbóreo, arenoso, boreal, boscoso, caducifolio, caudaloso, campero, coralino, costero, desértico, fluvial, forestal, frondoso, geológico, geomorfológico, glaciar, llano, marino, montañoso, natural, navegable, orográfico, paisajístico, panorámico, pecuario, rocoso, rupestre, rural, rupícola, salvaje, selvático, sinuoso, silvestre, volcánico	
	Manufacture	
	artesano	
	Material	
	biodegradable, calcáreo, calizo, granítico, kárstico, neumático, sintético	
	Motion-related	
	antideslizante, colgante, descendente, deslizante, insumergible, motriz, pendular, resbaladizo, sumergible	
	Physical property	
	ajustable, combinable, cristalino, desmontable, elástico, ergonómico, estanco, extraíble, hinchable, impermeable, inflable, inoxidable, plegable, telescópico, recio, reflectante, regulable, resistente, reutilizable, rígido, semirrígido, térmico, transpirable	
	Protection-related	
	protector	
Position/ Geographical location	*acuático, aéreo, delantero, horizontal, indoor, intermunicipal, insular, litoral, limítrofe, meridiano, meridional, noriental (nororiental), noroccidental, noroeste, outdoor, peninsular, subacuático, subterráneo, submarino, terrestre, vertical*	
Quantification	*doble, grupal, ilimitado, lleno, repleto, tándem*	
Temporal location	*prehispánico, milenario, vikingo*	
Temporality	*continuo, durable, estival, extraescolar, invernal, nocturno, otoñal, previo*	

Patrick Goethals

Ghent University
patrick.goethals@ugent.be

Jasper Degraeuwe

Ghent University
jasper.degraeuwe@ugent.be

4 Methodological advances in lexical pattern extraction: Examples from Spanish adventure tourism

Abstract: This chapter aims to contribute to the methodological innovation in the description of language use for specific purposes and the language of adventure tourism in particular. We describe techniques such as dependency parsing and semantic similarity calculation based on non-contextual word embeddings and transformer-based language models in order to show how these recent innovations or developments can lead to a more fruitful use of small and mid-size corpora, which are the typical use cases when studying languages for specific purposes. In these corpora, purely quantitative filters set too strict limitations on less frequently recurrent constructions. At the same time, we discuss the challenges posed by the new techniques, since they require adequate fine-tuning. Semantic similarity calculation in particular seems to offer important opportunities to explore the full richness of small and mid-size corpora such as the ADVENCOR corpus (Durán-Muñoz & Jiménez-Navarro, 2021), since it takes recurrent constructions as a starting point and enriches these data by identifying semantically similar constructions which by themselves do not pass the frequency thresholds. We will apply these methodological insights to verb constructions that are typical in the adventure tourism discourse.

Keywords: collocations, dependency parsing, pattern extraction, semantic similarity, transformers

1. Introduction

This chapter aims to contribute to the methodological innovation in the description of language use for specific purposes, and the language of adventure tourism in particular, by describing techniques that are relatively new or that have recently improved significantly, thanks to new evolutions in the field of

Natural Language Processing (NLP). We discuss dependency parsing for retrieving syntactic patterns and the use of word embeddings and transformer-based language models for calculating semantic similarity and clustering use patterns. We apply these techniques to the Spanish one-million-word web corpus of Adventure Tourism (ADVENCOR, Durán-Muñoz & Jiménez-Navarro, 2021), made available by the editors of this volume. We focus specifically on motion verbs, which show high degrees of keyness in this corpus (cf. Durán-Muñoz & Jiménez-Navarro, 2021).

One of the major challenges in the corpus-based description of language use is to achieve a sound balance between, on the one hand, collecting as many data as possible, and, on the other, economically identifying the most relevant data and reducing "noise" (e.g., Anthony, 2018). Traditionally, discourse analysts, lexicographers, or terminographers relied on small sets of well-selected sources. Within that approach, results were assumed to be relevant and accurate, but it was more difficult to provide evidence of exhaustiveness and representativeness. Current general language corpus techniques work in the opposite way (Davies & Parodi, 2022): by collecting large corpora and then analysing them quantitatively, the starting point is exhaustiveness and representativeness, and the challenge lies in achieving accuracy and relevance. Regarding accuracy, techniques are sought to reduce the number of false positives and false negatives and, to achieve satisfying levels of relevance, the complexity of the selected data must be reduced by incorporating hierarchies (the most relevant first), on the one hand, and clustering data (taking similar phenomena together), on the other. For both objectives mainly frequency-based methodologies are used. The specific challenge of working with corpora of language use for specific purposes is that these corpora will be smaller almost by definition, and that the frequency-based methodologies for working with big corpora should be complemented with other methodologies in order to obtain the most valuable return from the corpus analysis.

Within contemporary corpus analysis and the tools that support it (e.g., AntConc, WordSmith, and Sketch Engine [Kilgarriff et al., 2014]), a set of techniques have become standard over the past decade, enabling users to perform customised and more accurate queries. These techniques cover the computation of keyness, dispersion, or other frequency patterns in corpora, and allow performing partial preprocessing of the data in the form of lemmatisation and Part-of-Speech (PoS) analysis (Moreno Sandoval, 2022). Lemmatisation is an absolute must for a morphologically rich language like Spanish. Whereas English has less than 10 forms per verb, Spanish easily reaches 60 or more, resulting in frequency-based metrics quickly becoming unusable if non-lemmatised data are used. PoS tagging allows for morphosyntactic disambiguation of potentially ambiguous elements (e.g., Spanish *trabajador* as noun ['employee'] and adjective ['hard-working']) and also allows for the derivation of syntactic patterns with some degree of probability. In particular, in Sketch Engine, preprogrammed grammatical sketches are based on combinations of PoS tags. For example, the PoS combinations in column A are assumed to be indications of the syntactic relationships in column B.

Table 1. Rule-based inference of syntactic patterns based on PoS patterns

PoS Pattern	Hypothesis Syntactic Relation
<ARTICLE><NOUN><FINITE VERB>	subject – verb
<FINITE VERB><ARTICLE><NOUN>	verb – object
<NOUN><ADJECTIVE>	noun – modifier

Obviously, these are coarse patterns of analysis for which counterexamples are easy to find, especially in languages with a relative freedom in word order. For instance, *la monitora* in example (1), with pattern <FINITE VERB><ARTICLE><NOUN> is not the object, but the subject:

(1) *Nos acompañó la monitora Lourdes.* (ADVENCOR)

The default procedure to exclude such anomalous patterns from the analysis (and thus to avoid false positives) consists of filtering them out by incorporating relatively high frequency thresholds, and assuming that these erroneous patterns occur only sporadically and will therefore not pass the filter. An important caveat, however, is that this method works well for (very) large reference corpora, but it becomes much more difficult to handle in smaller specialised corpora. In this case, the hard-coded thresholds potentially lead to a large increase in false negatives, that is, relevant data that is lost. For instance, if we applied the frequently used threshold of more than three occurrences to the list of verb–object pairs (based on dependency parsing, Figure 1) in the ADVENCOR corpus, this would lead to the a priori exclusion of more than 90 % of the data.

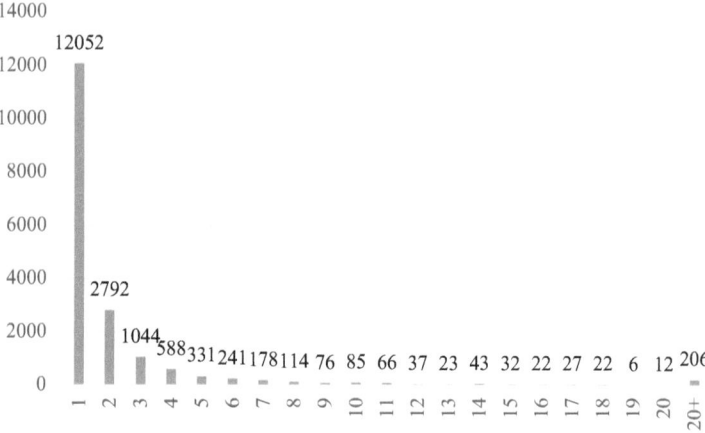

Figure 1. Frequency of verb–object pairs in ADVENCOR

Basically, this means that, in the case of smaller corpora, we would benefit from techniques that are able to overrule frequency thresholds.

Therefore, the purpose of this chapter is to take a look at some NLP techniques that are well established in specific NLP applications but have not yet found their way into the standard toolbox of descriptive linguists. In particular, we discuss dependency parsing and language modelling, and show how they can contribute to greater accuracy, on the one hand, and data clustering, on the other. At the same time, we also aim to already point out the complexity of the descriptive procedure when integrating these new tools. At a technical level, the analyses were carried out by making use of the preprocessing pipeline of Goethals et al. (2017). The new techniques will be made available in the near future in the SCAP platform[1] (scap.ugent.be).

---

1 The SCAP [Spanish Corpus Annotation Project] platform (scap.ugent.be, last accessed: 28/09/2023) was designed at Ghent University as a multi-purpose corpus tool, combining educational and research purposes. The platform provides access not only to corpus analysis tools, such as a preprocessing pipeline (PoS tagging, lemmatisation, and syntactic parsing) and quantitative measures (frequency, keyness), but also to didactic applications such as graded thematic wordlists, bilingual glossaries, or automatically generated fill-in exercises.

2. Pattern Extraction and (Universal) Dependency Parsing

In Table 1 we observe several examples of syntactic analysis that rely on rule-based pattern identification involving word order, PoS tagging, and function words such as articles or prepositions. This rule-based approach achieves a considerable degree of accuracy because language is largely composed of relatively simple grammatical patterns. However, more complex structures with long-distance dependencies or deviant word order are almost impossible to analyse using this rule-based approach. In fact, adding exception rules is time-consuming and may lead to the introduction of new errors, which makes it rarely, if ever, a completely successful operation. Automated syntactic parsing can provide a solution for this.

The most widely used contemporary technique for automatically analysing syntactic structure involves dependency parsing. Conceptually, dependency parsing defines the syntactic function of a word as a dependency relation to another word (the head) in the sentence. The word that functions as the initial head is labelled as the 'root'. Dependency parsing is a layer of preprocessing annotation that is not derived directly from the other layers such as lemmatisation, word order, or PoS tagging but is labelled by an independent machine learning algorithm. Especially since the introduction of neural machine learning, results have greatly improved. In Figure 2, for example, the Stanza parser (v.1.4.1) correctly identifies *la monitora* as the subject and not the object of the verb *acompañar*.

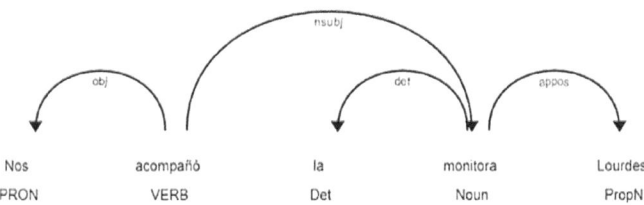

Figure 2. Dependency parsing (Stanza parser v.1.4.1, visualisation with displaCy module)

Table 2 presents the underlying data structure of the example in Figure 2. The dependency relation is defined by the attributes <deprel> (the relation type) and <head> (which refers to the index or position of the head of the dependency relation). This additional information makes it possible to define a corpus query that identifies the object or subject of the verb *acompañar* following the rule "identify the rows that have 'obj' or 'nsubj' as <deprel> attribute and whose <head> attribute refers to the <index> of the row with lemma attribute *acompañar*".

Table 2. Features of dependency parsing

index[2]	token	lemma	PoS	deprel	head
0	Nos	nos	PRON	obj	1
1	acompañó	acompañar	VERB	root	1
2	la	la	ARTDEF	det	3
3	monitora	monitora	NOUN	nsubj	1
4	Lourdes	Lourdes	PROPN	appos	3

Although, at present, dependency parsing is not yet part of the standard workflow facilitated by the best-known corpus tools, it is to be expected that it will be progressively introduced in the coming years (for a recent application in lexicography, see Orenha-Ottaiano et al., 2021). Moreover, the models are evolving quickly, are achieving higher and higher performance, and are covering more and more languages. The advantage over the PoS-based patterns described earlier is that, in principle, the parser is able to still find the right connections in more complex structures, for example, in case of word order discrepancies (example (1)) or when elements are more distant from each other. This is an important factor to reduce the number of false negatives. A good case in point are the verb–object and subject–verb instances: in about 20 % and 40 % of those instances, respectively, there are more than two words between the verb and the nominal element, making recognition via rule-based patterns a risky operation. In example (2), *descubrir* and *paisajes* are linked despite having another verb between them.

---

2   'Index' refers to the position of the word in the sentence. As a convention, counting starts at 0.

(2a) *Se proponen cinco actividades diferentes para <u>descubrir</u>, siempre andando, <u>paisajes</u>, flora, fauna, cultura.*[3]

(2b)

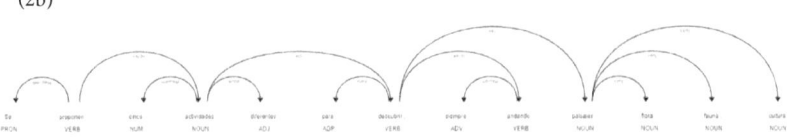

However, integrating dependency data into corpus queries is much less straightforward than in the case of lemmatisation or PoS data. A first important aspect to highlight in this regard is the recent trend to replace language-specific syntactic parsing by a cross-linguistic approach, propelled by the Universal Dependencies (UD) initiative (Nivre et al., 2016). This framework, which emerged in 2016, further developed previous proposals such as the universal Stanford dependencies (de Marneffe et al., 2014). The basic idea behind UD was that both the development of NLP tools and cross-linguistic analysis are hampered when each language is described with specific sets of PoS tags or dependency relations labels. Therefore, a set of universal labels was developed in order to generalise across typologically different languages. The constant expansion of the number of treebanks within the UD project[4] shows the vitality of this premise. Nevertheless, dependency parsing based on UD tags also leads to analyses that may deviate in specific ways from what would count as a standard analysis for a specific language. As the first applications of dependency parsing emerge within descriptive linguistics (Orenha-Ottaiano et al., 2021), it is important to be well aware of its specificities.

2.1. Challenges

Although working with UD parsing data bears great potential, it is also considerably more complex than working with lemma or PoS information because of: (1) its inherently relational nature (which makes queries become more complex), (2) the fact that UD tags do not always correspond with the default syntactic analysis in a specific language, and (3) noisy data caused by false negatives and false positives (i.e., errors generated by the algorithm). As an

---

3   All examples in this methodological section are extracted from the ADVENCOR corpus and minimally adapted in order to avoid unnecessary information in the parse trees. The examples are followed by the displaCy visualisation of the dependency parsing, and/or the underlying dependency features.

4   https://universaldependencies.org/ [Last accessed: 28/09/2023].

illustration, we discuss some specific phenomena related to two core dependency relations, namely object ('obj') and subject ('nsubj'), in order to show the type of query operations that are needed for an optimal design of the data retrieval (which should maximally avoid returning false negatives and false positives).

A first way to reduce the number of false negatives has to do with the (relatively) frequent cases where there are multiple coordinated objects of a verb (cf. example (3), where the verb *atravesar* has two direct objects, namely *puente* and *tirolina*). Such examples are very difficult to identify with rule-based scripts but become easily retrievable using dependency analysis. However, it should be noted that a relatively complex query would need to be defined to retrieve the examples since the coordinated objects themselves are labelled 'conj' and not 'obj', and the dependency relationship does not run directly between these objects and the verb, but through the first object. In essence, this means that a query solely based on the 'obj' tag would not identify the objects marked as 'conj'. In the <advenTourism> database, this involves more than 10 % of the cases, which would be a significant number of false negatives.

(3a) *Atravesarás un puente tibetano de 60m y una tirolina de 140m.*
(3b)

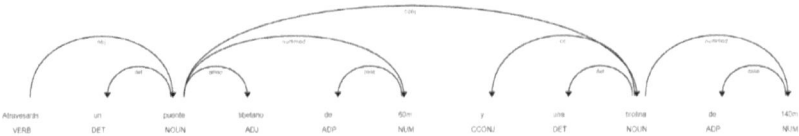

In example (4), we find another complex structure of highly relevant coordinated verb–object combinations:

(4a) *[Organizamos] excursiones para realizar senderismo, montañismo, escalada, barranquismo o cualquier otro deporte de aventura.*
(4b)

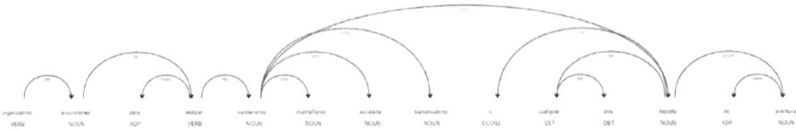

A second context where false negatives can be found is relative clauses, which, depending on the text genre, can be relatively frequent in Spanish. In example (5), it can be clearly seen how complex the underlying dependency structure is. To make the connection *recorrer – zona*, the query must identify a structure where the verb is in an 'acl' ('adjectival clause') relation with respect to a noun, and where there is a relative pronoun connected to the verb with the dependency relation 'obj'. Again, this represents a considerable number of examples (+ 5 % of the default verb–object relations) that would not be included through a standard query.

(5a) *La zona que recorreremos es preciosa.*
(5b)

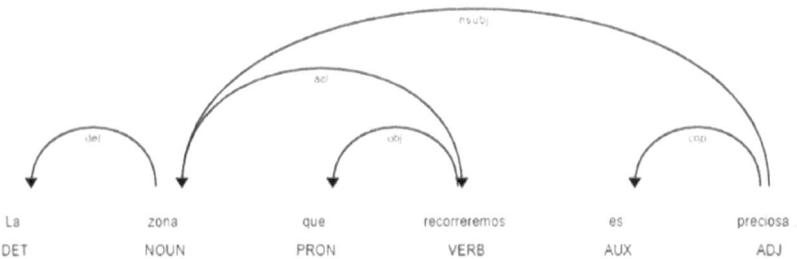

Third, there are the pronominal passive constructions with *se*, which are also interesting to be included in calculations that analyse which verbs are combined with which object–nouns. These constructions represent about 3.5 % of the number of verb–object constructions. Dependency information allows retrieving these examples, but it should again be emphasised that the query at hand would be of a rather complex nature, as it involves identifying a nominal element (*medidas*) that is in an 'nsubj' dependency relation with a verb (*adoptan*) that is, on its turn, also the head of a dependency relation with the passive marker *se* (coded as 'expl[etive]:pass[ive]').

(6a) *La escalada deportiva es un deporte completamente seguro, cuando <u>se adoptan las correspondientes medidas de seguridad.</u>*
(6b)

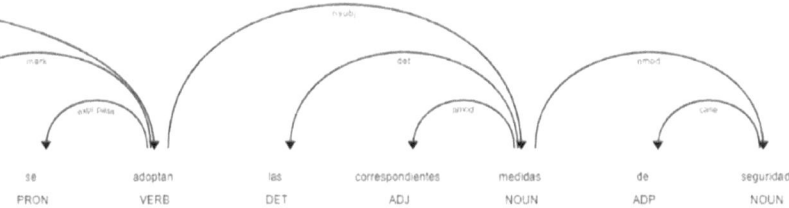

(6c)

index	token	lemma	PoS	deprel	head
10	se	se	PRON	expl:pass	11
11	adoptan	adoptar	VERB	advcl	5
12	las	el	DET	Det	14
13	correspondientes	correspondiente	ADJ	amod	14
14	medidas	medida	NOUN	nsubj	11
15	de	de	ADP	Case	16
16	seguridad	seguridad	NOUN	nmod	14

In examples (3)–(6), the focus was on merging different syntactic patterns to obtain as exhaustive a list as possible of combinations of verbs with a noun that would play the role of object in a prototypical singular active sentence. In this way, we maximally reduce the number of false negatives and make our dataset more robust when handling thresholds. In what follows, we also show some techniques to avoid false positives, again by taking into account the particularity of UD parsing, as well as by filtering out parsing errors.

A specific particularity of UD-based parsing concerns the subject relation in copula verbs. Unlike standard analysis for Spanish, UD does not connect the 'nsubj' dependency relation to the link verb (*ser*) but rather to the predicative adjective (*impermeable*) in example (7). Clearly, a query identifying possible subjects of verbs must filter out these cases by means of a check on the morphosyntactic category of the <head> of the dependency relation.

(7a) *Estas botas de montaña son totalmente impermeables.*
(7b)

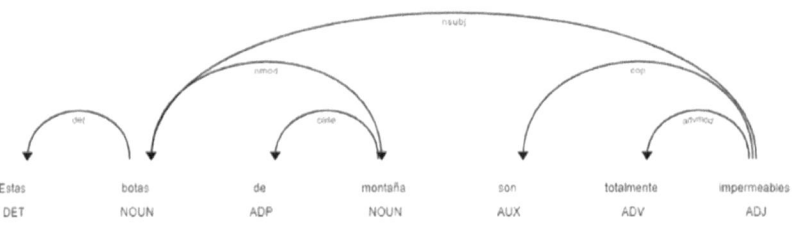

(7c)

index	token	lemma	PoS	deprel	head
0	*Estas*	*este*	DET	det	1
1	*botas*	*bota*	NOUN	nsubj	6
2	*de*	*de*	PREP	case	3
3	*montaña*	*montaña*	NOUN	nmod	1
4	*son*	*ser*	AUX	cop	6
5	*totalmente*	*totalmente*	ADV	advmod	6
6	*impermeables*	*impermeable*	ADJ	root	6

As mentioned earlier, automated annotation may also generate errors and false positives. Example (8) presents a common error in UD-based dependency parsing, with the argument *experiencia* being analysed as an 'obj' of the verb *disfrutar*. The standard analysis of Spanish only considers elements without a preposition or, under specific conditions, the preposition *a* as possible direct objects of the verb. A carefully crafted query can filter these examples by introducing a constraint that, in the case of the noun marked as 'obj', it must not be the head of a 'case' dependency, having as its lemma a preposition other than *a*. This is a recurrent and frequent deviation from standard analysis that generates a large number of false positives: if we limit ourselves to the prepositions *de*, *en*, *con*, and *por*, this already represents 12 % of cases initially defined as verb–object.

(8a) *Disfruta de una experiencia única.*
(8b)

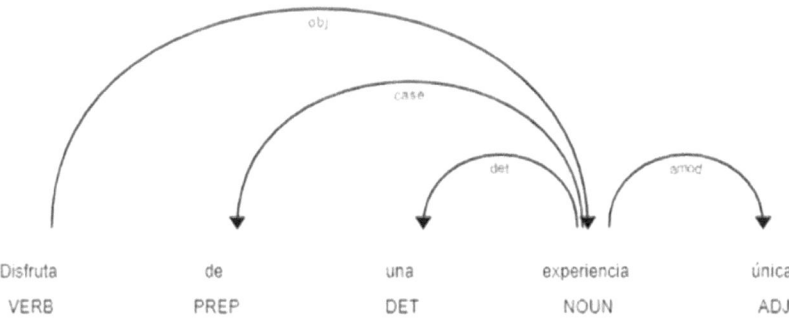

(8c)

index	token	lemma	PoS	deprel	head
0	*Disfruta*	*disfrutar*	VERB	root	0
1	*de*	*de*	ADP	case	3
2	*una*	*un*	DET	det	3
3	*experiencia*	*experiencia*	NOUN	obj	0
4	*única*	*único*	ADJ	amod	3

Another error that is relatively easy to detect is concordance conflicts between the verb and the nominal <subject> element. This is the case in example (9), where there is a conflict between the second person of the auxiliary verb and a third person noun subject. The information is derived from the morphology attribute that is also generated by the parser pipeline. A sample-based estimate indicates that 10 % of the subject relations identified by the current stanza parser can be discarded by adding a filter that checks the agreement requirements.

(9a) *El clásico <u>descenso</u> del Río Ara no te lo <u>puedes perder.</u>*
(9b)

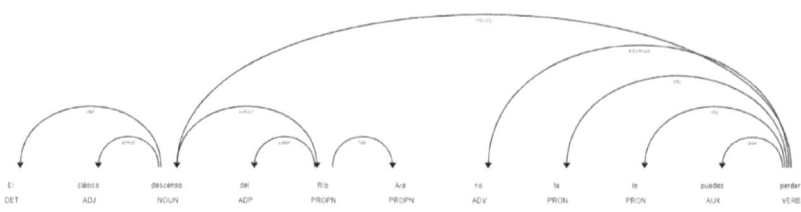

(9c)

index	token	lemma	PoS	deprel	head	morphology
0	*El*	*el*	DET	det	2	
1	*clásico*	*clásico*	ADJ	amod	2	
2	*descenso*	*descenso*	NOUN	nsubj	10	Number=Sing
3	*del*	*del*	ADP	case	4	
4	*Río*	*Río*	PROPN	nmod	2	
5	*Ara*	*Ara*	PROPN	flat	4	
6	*no*	*no*	ADV	advmod	10	

index	token	lemma	PoS	deprel	head	morphology
7	te	tú	PRON	iobj	10	
8	lo	él	PRON	obj	10	
9	puedes	poder	AUX	aux	10	Number=Sing|Person=2
10	perder	perder	VERB	root	10	VerbForm=Inf

As a conclusion to the previous discussion of dependency parsing, we would like to point out that, on the one hand, dependency parsing will likely become common practice in the coming years for the extraction of patterns from corpus data. On the other hand, however, it should be noted that it becomes particularly important to carefully define queries and use careful methods to limit the number of false negatives and false positives. Our estimates show that appropriate filters affect the annotation of about 30 % of the subject or object relationships.

Once the extraction queries have been applied to generate lists of potentially relevant patterns, the resulting lists have to be sorted in order to prioritise the most relevant or typical data, and to link similar data in a relevant and preferably semantically motivated way. Hierarchisation is covered briefly in Section 3, and clustering is discussed in more detail in Section 4.

3. Hierarchisation of Patterns

The main challenge in processing automatically generated corpus output is designing a methodology that allows exploiting the richness of the data without spending considerable time going over a very large number of constructions and examples. Table 3 shows the result of a grouping and hierarchisation methodology with the verb *saltar* in the ADVENCOR corpus. The following data are incorporated:

1. Parsing relations 'object' and 'oblique' ('collocate').
2. Frequency of the combination *saltar* – collocate (FILTER 1 = threshold +1) ('freq').
3. Keyness value of the collocate (typicality score of the collocate in this corpus) (FILTER 2 = sorted in descending order) ('key').
4. Delta-value (typicality score of the collocate for *saltar* in this corpus) ('delta').
5. The function words that link the nucleus and the collocate (e.g., prepositions and articles) ('link').
6. Extra dependency relations associated with the nucleus or the collocate ('pattern').

7. Example contexts ('context').

The methodology reduces the total number of +100 examples into 23 grouped patterns, which are sorted according to the keyness value of the collocate. Although this methodology is relatively well known and accessible in other formats, we draw attention to the specific feature of adding information based on the analysis of the extra dependency relations. Although this generates some noise (*saltar al agua de manera*), it also yields combinations such as *saltar de roca en roca, saltar en pozas profundas, saltar del puente en una caída libre, saltar sin objetos en los bolsillos*, or *saltar al vacío en paracaídas*, which are extremely useful in order to show the meaning of the word in context.

Table 3. Clustering of patterns based on dependency parsing (frequency threshold = +1, sorted by descending keyness of the <obj> collocate)

SALTAR						
Collocate	Freq	Key	Delta	Link	Pattern	Context
tirolina	2	0.95	0.05	en artdef	*saltar en la tirolina*	tomar la decisión de saltar en la tirolina, nadie ha podido dejar de mirar hacia abajo
puenting	3	0.89	0.17	en	*saltar en puenting*	Todas las empresas para saltar en puenting en Barcelona.
poza	21	0.74	0.58	A	*saltar a pozas*	descender por toboganes, saltar a pozas o pasar por desfiladeros estrechísimos en oposición.
poza	21	0.74	0.58	A	*saltar a pozas profundas*	dejarnos deslizar por toboganes, saltar a pozas profundas o pasar por estrechos desfiladeros en oposición.
poza	21	0.74	0.58	a artdef	*saltarse a la poza*	para cuando no se puede saltar a la poza, ya sea por poca profundidad o por nuestro
paracaídas	6	0.7	0.29	en	*saltar en paracaídas*	La sensación de saltar al vacío en paracaídas a 4.000 metros es única y vale la pena
seguridad	2	0.56	0.04	con	*saltar con seguridad*	donde nuestros guías nos mostrarán cómo saltar con seguridad.

(*continued*)

Table 3. Continued

				SALTAR		
Collocate	Freq	Key	Delta	Link	Pattern	Context
puente	9	0.56	0.19	de artind	saltar de un puente	llegando a una cascada o saltar de un puente.
puente	9	0.56	0.19	desde artind	saltar desde un puente	Saltar al vacío desde un puente ofrece sensaciones únicas.
agua	16	0.52	0.38	a artdef	saltar al agua	tendrás la oportunidad de saltar al agua y hacer snorkel con coloridos peces de arrecife,
agua	16	0.52	0.38	a artdef	saltar al agua de manera	mientras el guía nos ayuda a rapelar o saltar al agua de manera segura.
roca	4	0.52	0.09	de	saltar de roca en roca	vadear, nadar, saltar de roca en roca, saltar de rocas al agua,
extremo	2	0.5	0.06	desde artind	desde un extremo del puente saltar	Desde un extremo del puente se salta, la cuerda, que
altura	3	0.5	0.06	desde diferente	saltar desde diferentes alturas	tendremos que saltar desde diferentes alturas, normalmente se comienza por saltos de unos 2
caída	3	0.49	0.11	en artind	desde el puente saltar en una caída libre	desde el puente saltamos al vacío en una caída libre y luego en movimientos tipo columpio quedamos suspendidos
día	2	0.44	0.03	qu	saltar cualquier día de la semana	por la zona de mayo a septiembre podrás saltar cualquier día de la semana, el resto del año,
manera	3	0.4	0.06	de	saltar de manera segura	mientras el guía nos ayuda a rapelar o saltar al agua de manera segura.
avión	4	0.38	0.18	desde artind	saltar desde un avión	Si siempre has soñado con saltar desde un avión y disfrutar al máximo

(continued)

SALTAR					
Collocate	Freq	Key	Delta	Link	Pattern	Context
ocasión	2	0.38	0.04	en qu	en algunas ocasiones saltar	En algunas ocasiones tendremos que saltar para alcanzar el cauce del río
objeto	3	0.37	0.14	sin	saltar sin objetos en los bolsillos	Hay que recordar que se debe saltar sin objetos en los bolsillos o que puedan desprenderse.
cabeza	3	0.23	0.1	de	saltar de cabeza	Si se desea saltar de cabeza, deberá hacerse con decisión impulsándose con fuerza hacia
vacío	21	0.22	0.61	a artdef	saltar al vacío	a través de un arnés y saltáis juntos al vacío… ¡4 minutos de caída antes de aterrizar
vacío	21	0.22	0.61	a artdef	saltar al vacío en paracaídas	La sensación de saltar al vacío en paracaídas a 4.000 metros es única y vale

4. Semantic Clustering of Patterns

Conventional corpus queries combining word, PoS, or lemma information yield lists of examples which human coders then further analyse in order to identify categories of examples. As we have seen in Section 3, coders can make use of corpus analysis techniques to group and sort examples according to formal criteria such as dependency relations, or according to quantitative measurements such as frequency or keyness. In what follows we discuss new, complementary techniques that cluster items on semantic grounds. As a starting point, we show examples (10) and (11) in which the corpus query output still has to be clustered: this cannot only be done manually but also with the techniques that are explained in Section 4.1. In Sections 4.2. and 4.3. we then apply the techniques to the examples.

In example (10a), our starting point is the construction *escalar una montaña*, and the question is which other verbs take *una montaña* as an object and could be grouped together with *escalar*. To identify these semantically related words, a human coder would have no other option than reading the full list of 41 verbs.[5] In Section 4.2. we see that word embeddings will help us to identify these target items.

(10a) *escalar una montaña* > other verbs which take *montaña* as object?

> *albergar, alternar, amar,* **ascender,** *atraer, atravesar, avistar,* **bajar,** *conocer, considerar, contar, contemplar, contrastar, cortar, desafiar, descubrir, disfrutar, dominar, ejercer, encontrar, esculpir, explorar, hacer, implicar, incluir, inundar, llevar, mover, participar, pasar, pedir, poseer, practicar, preferir, recorrer, rodar,* **subir,** *tener,* **trepar,** *usar, ver*

Similarly, in Table 4 we see the outcome of a query of the <obj> dependency relations with the verb *pasar*, sorted according to decreasing degree of keyness of the <obj> collocate. As was also the case in example (10), human coders would again have to manually detect semantic relationships between the items, for instance, by clustering the constructions in bold around the meaning /*pasar cierto tiempo*/ ('to spend some time').

Table 4. <obj> dependency relations with the verb *pasar*

pasar rápeles	**pasar una jornada**	pasar un fin	pasar calor
pasar las localidades	pasar el puente	pasar una estancia	pasar la crisis
pasar esta actividad	pasar ramio	pasarse la parte	pasar frío
pasar pasamanos	pasar una cuerda	pasar la oportunidad	pasar la pelota
pasar nuestros circuitos	pasar la información	pasar a los coches	pasar ningún miedo
pasar los participantes	pasar el link	pasar un saludo	pasar al lado
pasar el río	pasar la roca	**pasar un tiempo**	pasar a su contrario
pasar la montaña	pasar una bayeta	**pasar estas fiestas**	**pasar un rato**
pasar unos días	pasar tus vacaciones	pasar la ocasión	pasar el testigo
pasar al nivel	pasar obstáculos	pasar un paso	pasar un trago
pasar al valle	**pasar un verano**	pasar el resto	pasar la gota
pasar una balsa	**pasar el invierno**	**pasar la noche**	pasar el hilo
pasar a mi novia	pasar medidas	pasar momentos	pasar a mi hermano
pasar un enlace	**pasar /cifra/ horas**	pasar una mañana	
pasar las cascadas	pasar el rincón	pasar una tarde	
pasar tramos	**pasar un día**	pasar /cifra/ túneles	

---

5   Verbs are alphabetically sorted; related target verbs expressing the idea of going up/down are marked manually in bold.

4.1. Word embeddings and transformer-based language models

Word embeddings (since 2013; Mikolov et al., 2013) and transformer-based language models (since 2017; Vaswani et al., 2017) have been revolutionising NLP tasks such as machine translation, sentiment detection, or, most recently, text generation (GPT3, ChatGPT). Non-contextual or static word embeddings represent each word form as a vector, that is, a fixed list of values, that is inferred from the distribution of the word form in large corpora. Drawing upon the assumption that distribution is a quantifiable proxy for word meaning and word function, word embeddings can be used to link semantically similar word forms (it is important to note that word embeddings are associated with word forms and not with lemmas). The assumption is, of course, that semantically similar words exhibit similar distributions and are thus represented by similar vectors. A quick look at an older and relatively simple 50-item vector visualisation of the nouns *empresa*, *compañía*, and *comida* shows that the two most similar vectors are those of *empresa* and *compañía* (current vectors would use +300 dimensions, which makes visual comparison more difficult).

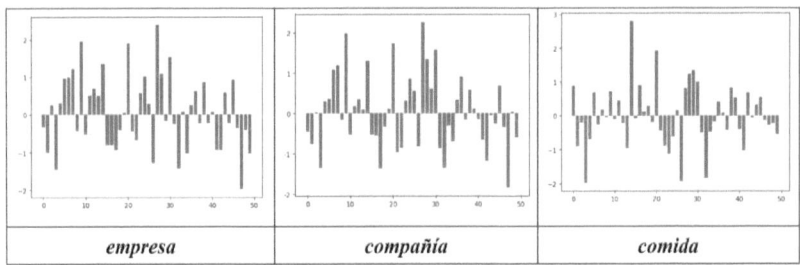

Figure 3. Non-contextual word embedding representations

The vectors of the words can also be projected in a two-dimensional space, which visually shows that semantically related words are grouped together. In Figure 4, we can observe some words related to two different activities, namely parachute jumping and canyoning. Interestingly, words related to the former activity are grouped on the left, words related to the latter activity appear on the right, and the shared verb *saltar* ('to jump') is situated in between.

Methodological advances in lexical pattern extraction 103

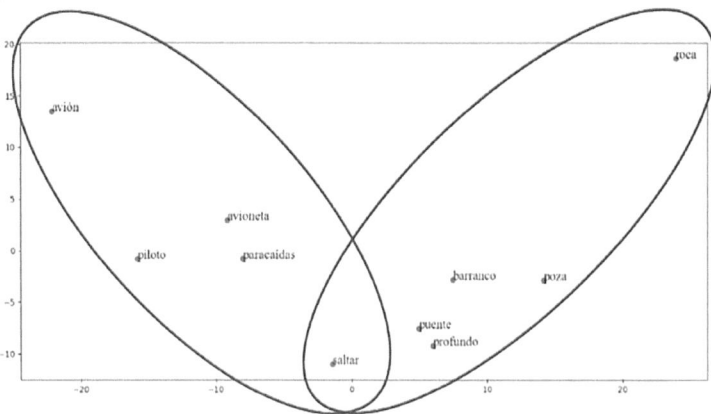

Figure 4. Two-dimensional projection of non-contextual word embedding vector representations

The similarity between two words (or word vectors) is typically calculated via cosine similarity, which is the cosine of the angle between the two vectors. Cosine similarity calculations yield scores between -1 and +1, with +1 representing parallel vectors and thus maximal similarity. In other words, by comparing cosine similarity, we can order words according to their similarity to a reference word. This is done in Table 5 using the word *avión* as a reference word. We can clearly see that the words from the left side of Figure 4 (i.e., *avioneta, paracaídas, piloto*) indeed achieve the highest cosine similarity scores.

Table 5. Cosine similarity calculation based on non-contextual word embeddings. Reference point *avión*

avión	
avioneta	0.809
paracaídas	0.602
piloto	0.505
puente	0.298
barranco	0.218
saltar	0.205
roca	0.116
poza	0.053
profundo	0.023

However, non-contextual word embeddings also face severe limitations, the most important one being that a word form is represented by the same vector in every context. In the case of polysemous word forms such as *compañía* (meaning 'firm' in example (11) or 'companionship' in example (12)), or *escala* (meaning 'scale' as noun or 's/he climbs' as verb), this means that they are always represented by the same vector. As a consequence, word forms that are ambiguous between PoS categories yield very deviant results when they are used in similarity calculations.

(11) *Billete de avión en vuelo regular con la compañía aérea Royal Jordania* (ADVENCOR).
(12) *El descenso en canoa rafting lo hacemos siempre en compañía de un monitor* (ADVENCOR).

From 2017 onwards, transformer-based language models have added an extra and crucial dimension to non-contextual word embeddings because they create a representation that takes into account the word form's own profile as well as the profile and position of its context words. As a result, the transformer representation of example (11) can be expected to take an economic "touch" and, for example, be closer to the word *empresa* than example (12). Contrary to non-contextual word embeddings, contextual embeddings generated by transformer models thus allow calculating semantic similarity between two different word forms in context, and to disambiguate uses of the same polysemous word (Degraeuwe & Goethals, 2022). The most recent and comprehensive language model for Spanish is RoBERTa-large (Gutiérrez-Fandiño et al., 2021; Liu et al., 2019), which we use in the application described in Section 4.2.

4.2. Application of non-contextual word embeddings

As a first application we will sort the list of verbs that take *montaña* as object, in order to search for verbs that would belong to the same semantic domain as *escalar* in *escalar una montaña*. In example (10b), the verbs are ordered according to descending frequency in the corpus. As can be expected, this does not help much to identify in a faster way the related verbs. In example (10c), we calculated the average vector similarity between the infinitive forms, third person simple present, and third person simple past, meaning that we compare, for instance, *escalar-ascender / escala-asciende / escaló-ascendió* and calculate the average. Thanks to this method, the target candidates move closer to the top positions. But the best results are clearly to be found in example (10d), where we fine-tuned the comparison by leaving out those items that are ambiguous with respect to the PoS status (e.g., *escala* or *baja*). With the last methodology, the four targets occur within the top-5.

(10b) *disfrutar,* **ascender,** *conocer,* **subir,** *recorrer,* **trepar,** *ver, atravesar, contemplar, preferir, amar, descubrir, encontrar, explorar, incluir, albergar, alternar, atraer, avistar,* **bajar,** *considerar, contar, contrastar, cortar, desafiar, dominar, ejercer, esculpir, hacer, implicar, inundar, llevar, mover, participar, pasar, pedir, poseer, practicar, rodar, tener, usar*

(10c) **ascender, bajar,** *recorrer, dominar, desafiar, encontrar,* **subir,** *contrastar, implicar,* **trepar,** *mover, atravesar, cortar, considerar, descubrir, poseer, explorar, contar, practicar, contemplar, llevar, incluir, atraer, alternar, hacer, conocer, rodar, albergar, pasar, ejercer, participar, usar, inundar, preferir, esculpir, ver, disfrutar, tener, pedir, amar, avistar*

(10d) **ascender, bajar,** *dominar,* **subir, trepar,** *mover, desafiar, recorrer, encontrar, cortar, atravesar, pasar, descubrir, implicar, llevar, atraer, contrastar, hacer, poseer, ver, considerar, contar, explorar, incluir, conocer, ejercer, practicar, usar, rodar, contemplar, participar, albergar, inundar, alternar, preferir, esculpir, disfrutar, amar, pedir, avistar, tener*

Of course, the effect of efficient sorting becomes more important when the group of items to be considered is longer. As an example, we applied the same procedure to the full list of 323 verbs that share at some point the same object as *escalar*. We identified five clear motion verbs in the list and, as can be seen in example (13), these verbs all occur in the top-11 when the fine-tuned non-contextual semantic similarity method is used.

(13) **descender, ascender,** *alcanzar,* **bajar,** *sumergir, dominar,* **subir,** *superar, llegar, aumentar,* **trepar,** *embarcar, coronar, desfigurar, incrementar, reducir, desmontar, levantar, distar...* [323 verbs in total]

4.3. Application: Transformer-based language models

We now show what can be the contribution of the technique of the transformer-based language model by applying it to the previously mentioned list of <obj> patterns with the verb *pasar* (Table 4). Let us suppose that the human coder identifies *pasar una jornada* as potentially relevant and wants to know which nouns display a similar semantic profile.

(14) *Ven con nosotros a <u>pasar una jornada</u> de deportes de aventura mientras recorremos la Vía Ferrata del Torreón y el barranco [...]* (ADVENCOR).

Instead of manually going over all the sentences or patterns, we apply the language model by taking the word *jornada* as a reference point within this specific sentence and calculating the semantic similarity of the lexical elements occupying the same <obj> position in the rest of the (more than 200) sentences containing *pasar* combined with an object. The sentences are sorted in descending degree of similarity, and we extract each time the first and thus most similar sentence for a given lemma. Table 6 shows the most similar 15 lemmas according to the automatic classification.[6]

Table 6. Cosine similarity calculus based on transformer language model RoBERTa

Lemma	Example	Simil score
día	¿Quieres pasar un día de aventura y emoción en la naturaleza?	0.87
mañana	que buscas es una buena manera de pasar una encantadora mañana en familia o en grupo, tanto si quieres	0.84
tarde	cuya intención no es otra que la de pasar una tarde en familia.	0.80
actividad	No puedes dejar pasar esta actividad donde subiremos por paredes verticales.	0.74
noche	Si quieres pasar una noche loca de verdad, elige un espectáculo Drag-queen y	0.74
rato	La mejor forma para pasar un rato inolvidable con la familia y amigos, en un	0.73
momento	el complemento ideal para buceadores y para pasar grandes momentos en el mar o en el lago.	0.71
tiempo	por un río de aguas bravas y pasar así un tiempo emocionante y divertido al mismo tiempo.	0.70
fin	en formarte como buzo profesional o simplemente quieres pasar un fin de semana completo sumergido en las profundidades de las	0.70
vacación	open kayak o descenso de cañones para pasar unas vacaciones multiaventura inolvidables en el Pirineo aragonés.	0.70
puente	La verdad que vengo de pasar un puente inolvidable, me ha gustado todo, la gente	0.66
estancia	Molinos, hoy abandonado, donde podemos pasar una agradable estancia en el albergue de los molinos y disfrutar de	0.66
oportunidad	No dejes pasar esta oportunidad.	0.64
circuito	escenarios diferentes con 150 bolas, consigue pasar todos nuestros circuitos de aventura llenos de emoción con rocódromos, lianas	0.63
verano	genial para que los peques de la casa pasen un verano inolvidable.	0.62
…	+200 sentences	

The result is strikingly accurate and provides a great help in clustering the data, as constructions semantically related to *pasar una jornada* come first in the

---

6   Items marked in bold would be the items that are manually selected.

list. The human annotator can now confirm the choice much faster and deduce a pattern along the lines of example (15):

(15) *pasar cierto tiempo (jornada, día, mañana, tarde, noche, un fin de semana, un puente…)* (ADVENCOR).

The example with *puente* is particularly interesting because we see that *puente* indeed refers to a time period ('holiday weekend') and not to a physical bridge. Examples with the latter meaning obtain lower similarity scores.

Finally, it is important to emphasise that the semantic clustering method allows lowering or even omitting hard-coded frequency thresholds. For example, the automatised retrieval method only found one instance of *pasar un verano*, and thus the construction would not pass any frequency threshold. In the new methodology we would start by considering the most frequent combinations, and then recover data from the patterns that by themselves do not pass the frequency threshold but are relevant on the basis of their semantic similarity with the frequent patterns. In summary, semantic similarity hierarchisation is especially well suited for small or mid-size corpora where, otherwise, too much relevant information is filtered out by frequency thresholds.

4.4. Visualisation of clustering: Graph Theory

As a final application, we would like to mention Graph Theory (Python networkx module), which can be integrated in the workflow in order to visualise networks of items and the strength of the connections between them. This network can be visualised as a graph, which helps to understand how the words interact with each other. As an example, Figure 5 displays the distribution of nouns in an <obj> dependency relation with the verbs *escalar, ascender, subir,* and *trepar* (frequency threshold +1). This kind of automatically generated figures will also help linguists in the bottom-up approach to discourse and language use.

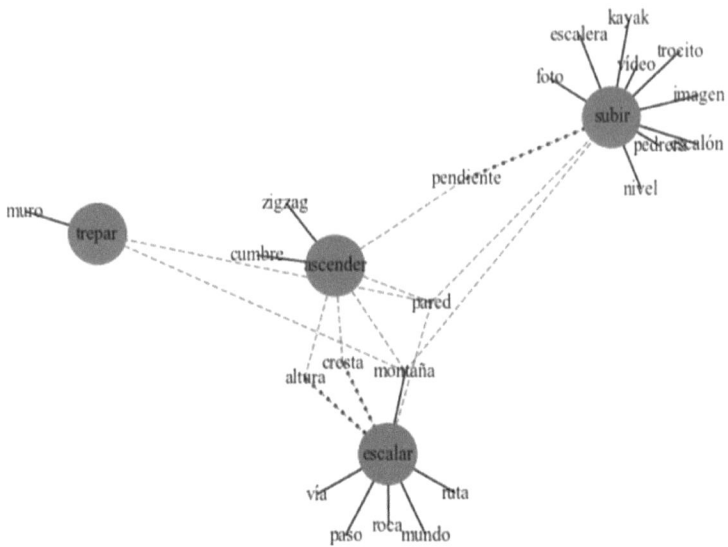

Figure 5. Graph representation of the verb–object distribution with the related verbs *escalar*, *ascender*, *subir*, and *trepar*

5. Conclusions

In this contribution we explained recent corpus analysis techniques and applied them to a small selection of motion verbs such as *saltar* and *escalar*, belonging to the top-20 verbs with the highest keyness value in the ADVENCOR corpus (cf. Degraeuwe & Goethals, 2020, for details about keyness calculation). One of the main challenges of corpus research on languages for specific purposes is the a priori size limitations of the corpora.

Small- to mid-size corpora (1 to 5 million words), such as ADVENCOR, offer specific advantages and disadvantages: their relatively small size guarantees their inherent coherence, and hence their relevance for the specific discourse that is studied. Indeed, they are ideal sources for the development of dictionaries or language learning materials, or for applying discourse analytic techniques. At a methodological level, however, a dilemma arises: on the one hand, they are too large to be studied exhaustively, but, on the other hand, the default techniques that are used to select data from large corpora are too coarse-grained and risk to filter out too much data. Therefore, we want to advocate that an extra set of NLP-based methodological tools can be introduced.

First, we showed that it is helpful to fine-tune data selection by designing queries that make use of dependency parsing instead of PoS tagging and lemmatisation only, although at the same time we emphasised the importance of reducing the number of false positives and false negatives by incorporating extra checks in the queries.

Secondly, we explained different techniques to cluster data, either based on extended dependency information, or by using the most recent NLP developments of word embeddings and transformer-based language models. The advantage of the latter methods is that they succeed very well in bringing the semantically most relevant data to the top of the hierarchy and can therefore override frequency thresholds.

We hope that we have raised interest in some new NLP-inspired techniques, which in our view can be very helpful for future lexicographic work on adventure tourism such as the *DicoAdventure* project, and, more broadly speaking, can further stimulate fine-grained empirical research based on corpora that are often inherently limited in size because of their specialised character.

References

Anthony, L. (2018). Visualisation in corpus-based discourse studies. In C. Taylor & A. Marchi (Eds.), *Corpus approaches to discourse* (pp. 197–224). Routledge.

Davies, M., & Parodi, G. (2022). Constitución de corpus crecientes del español. In G. Parodi, P. Cantos Gómez, & C. Howe (Eds.), *Lingüística de corpus en español* (pp. 13–32). Routledge.

Degraeuwe, J., & Goethals, P. (2020). La selección temática del vocabulario para fines didácticos: Evaluación de un acercamiento cuantitativo. *Revista de Lingüística y Lenguas Aplicadas*, *15*, 1–14. https://doi.org/10.4995/rlyla.2020.11969

Degraeuwe, J., & Goethals, P. (2022). Interactive word sense disambiguation in foreign language learning. In D. Alfter, E. Volodina, T. François, P. Desmet, F. Cornillie, A. Jönsson, & E. Rennes (Eds.), *Proceedings of the 11th Workshop on Natural Language Processing for Computer-Assisted Language Learning (NLP4CALL 2022)* (Vol. 190) (pp. 46–54). Linköping Electronic Conference Proceedings. https://doi.org/10.3384/ecp190005

Durán-Muñoz, I., & Jiménez-Navarro, E. L. (2021). Colocaciones verbales en el turismo de aventura: Estudio contrastivo inglés-español. In G. Corpas Pastor, M.ª R. Bautista Zambrana, & C. M. Hidalgo-Ternero (Eds.), *Sistemas fraseológicos en contraste: Enfoques computacionales y de corpus* (pp. 121–142). Comares.

Goethals, P., Lefever, E., & Macken, L. (2017). SCAP_TT: Tagging and lemmatising Spanish tourism discourse, and beyond. *Ibérica*, *33*, 279–288.

Gutiérrez-Fandiño, A., Armengol-Estapé, J., Pàmies, M., Llop-Palao, J., Silveira-Ocampo, J., Carrino, C. P., Gonzalez-Agirre, A., Armentano-Oller, C., Rodriguez-Penagos, C., & Villegas, M. (2021). Spanish language models. *arXiv preprint arXiv:2107.07253*.

Kilgarriff, A., Baisa, V., Bušta, J., Jakubíček, M., Kovář, V., Michelfeit, J., Rychlý, P., & Suchomel, V. (2014). The Sketch Engine: Ten years on. *Lexicography*, *1*(1), 7–36.

Liu, Y., Ott, M., Goyal, N., Du, J., Joshi, M., Chen, D., Levy, O., Lewis, M., Zettlemoyer, L., & Stoyanov, V. (2019). Roberta: A robustly optimized bert pretraining approach. *arXiv preprint arXiv:1907.11692*.

de Marneffe, M.-C., Dozat, T., Silveira, N., Haverinen, K., Ginter, F., Nivre, J., & Manning, C. D. (2014). Universal Stanford dependencies: A cross-linguistic typology. In N. Calzolari et al. (Eds.), *Proceedings of the Ninth International Conference on Language Resources and Evaluation (LREC'14)* (pp. 4585–4592). European Language Resources Association (ELRA). http://www.lrec-conf.org/proceedings/lrec2014/pdf/1062_Paper.pdf

Mikolov, T., Chen, K., Corrado, G., & Dean, J. (2013). Efficient estimation of word representations in vector space. *arXiv preprint arXiv:1301.3781*.

Moreno Sandoval, A. (2022). Etiquetadores morfosintácticos para corpus en español. In G. Parodi, P. Cantos Gómez, & C. Howe (Eds.), *Lingüística de corpus en español* (pp. 404–418). Routledge.

Nivre, J., de Marneffe, M. C., Ginter, F., Goldberg, Y., Hajic, J., Manning, C., McDonald, R., Petrov, S., Pyysalo, S., Silveira, N., Tsarfaty, R., & Zeman, D. (2016). Universal dependencies v1: A multilingual treebank collection. In N. Calzolari et al. (Eds.), *Proceedings of the Tenth International Conference on Language Resources and Evaluation (LREC'16)* (pp. 1659–1666). European Language Resources Association (ELRA). https://aclanthology.org/L16-1262.pdf

Orenha-Ottaiano, A., Garcia, M., Eugênia, M., de Oliveira Silva, O., L'Homme, M.-C., Ramos, M., Valêncio, C., & Tenório, W. (2021). Corpus-based methodology for an online multilingual collocations dictionary: First steps. In I. Kosem, M. Cukr, M. Jakubíček, J. Kallas, S. Krek, & C. Tiberius (Eds.), *Electronic lexicography in the 21st century. Proceedings of the eLex 2021 conference* (pp. 1–28). Lexical Computing CZ, s.r.o. https://elex.link/elex2021/wp-content/uploads/2021/08/eLex_2021_01_pp1-28.pdf

Vaswani, A., Shazeer, N., Parmar, N., Uszkoreit, J., Jones, L., Gomez, A., Kaiser, L., & Polosukhin, I. (2017). Attention is all you need. *Advances in Neural Information Processing Systems*, *30*. https://doi.org/10.48550/arXiv.1706.03762

Eduardo José Jacinto García

Universidad de Córdoba
ejacinto@uco.es

5 The argument structure of motion verbs in Spanish: A methodological proposal applied to *DicoAdventure*

Abstract: This chapter describes the methodology employed in the terminological resource *DicoAdventure* (Durán-Muñoz, 2021; Durán-Muñoz & L'Homme, 2020) to represent the argument structure of Spanish motion verbs in the language of adventure tourism. Its main purpose is to provide an exhaustive description of the differences and similarities of the selected verbs regarding their argument structure and the participants involved in them, and categorise them according to their syntactic-semantic behaviour. To do so, we rely on a well-established classification (Morimoto, 2001) which distinguishes between inherently directed motion verbs (IDMV), that is, those that express displacement with a specific trajectory (e.g., *acercarse* ['approach'], *descender* ['descend']), and manner-of-motion verbs (MMV), which show the existence of displacement but do not express the type of trajectory implied (e.g., *caminar* ['walk'], *nadar* ['swim']). In addition, given their presence in texts related to adventure tourism, we also analysed action verbs which, strictly speaking, are not motion verbs and, therefore, are not studied by Morimoto (2001), but which do imply motion and denote the control of a vehicle, such as *pedalear* ['pedal'] or *palear* ['paddle'].

Keywords: action verb, adventure tourism, argument structure, inherently directed motion verb, manner-of-motion verb, motion verb

1. Introduction

The present study aims to analyse the different argument structures of a set of motion verbs related to adventure tourism, more specifically, those that are lexicalised in the *DicoAdventure* dictionary[1] at the time of this study, and categorise them according to the proposal made by Morimoto (2001), who bases her studies on the Conceptual Semantics developed by Jackendoff (1990, 1992). To do so, this research also builds upon the following previous

---

1 The *DicoAdventure* dictionary is a specialised resource developed within the framework of the *DicoAdventure* project, led by Isabel Durán-Muñoz, and freely available at http://olst.ling.umontreal.ca/dicoadventure/ [Last accessed: 28/09/2023].

works: (1) Durán-Muñoz (2016), which lays the theoretical and methodological foundations to elaborate *DicoAdventure* and follows the principles outlined by Frame Semantics (Fillmore, 1977, 1982, 1985); (2) L'Homme (2007), which presents the advantages of adapting Explanatory Combinatorial Lexicology not only to lexicographical repertoires of the general language but also to terminological resources; and (3) Durán-Muñoz and L'Homme (2020), which details how both approaches can be combined, paying attention to motion verbs in the language of adventure tourism. Moreover, we take into account the PhD thesis by Jiménez-Navarro (2020), which analyses the collocations of motion verbs in this specialised language.

As mentioned above, *DicoAdventure* is based on the principles postulated by Frame Semantics (Fillmore, 1977, 1982, 1985), a cognitive approach which, according to L'Homme (2020), suggests many parallels with Explanatory Combinatorial Lexicology; in fact, these two approaches can complement each other, as demonstrated by the terminological resources *DiCoEnviro*,[2] and *DiCoInfo*,[3] developed in the Observatoire de linguistique Sens-Texte (OLST) at the University of Montreal and led by Marie-Claude L'Homme:

> However, Frame Semantics [FS] and Explanatory Combinatorial Lexicology [ECL] do share some methodological principles. The 'lexical unit' [LU] is defined according to very similar criteria. Both frameworks account, albeit differently, for the interface between the semantic and the syntactic properties of LUs. The similarities between the frameworks are more easily perceptible in the lexical resources which are based on them. It is our contention that FS and ECL complement each other. ECL allows us to provide a very detailed picture of the lexico-semantic properties of terms. FS goes slightly beyond and allows us to connect terms and their properties to broader situations. This connection informs us on how events, properties and even entities interact in a given field of knowledge. (p. 50)

Frame Semantics and its best-known application, FrameNet,[4] start from the idea that words evoke in our minds frames or scenarios that are fixed in memory and display prototypical characteristics. Their nature is anchored in human knowledge from experience. In both theories (i.e., Frame Semantics and Explanatory Combinatorial Lexicology) predicative units, in this case verbs, are

---

2 http://olst.ling.umontreal.ca/cgi-bin/dicoenviro/search-enviro.cgi?ui=en&mode= terme&lang=fr&prec= exact &equi=1&rq= [Last accessed: 28/09/2023].
3 http://olst.ling.umontreal.ca/cgi-bin/dicoinfo/search.cgi [Last accessed: 28/09/2023].
4 https://framenet.icsi.berkeley.edu/ [Last accessed: 28/09/2023]. FrameNet is a digital resource created in 1997 which applies the principles of Frame Semantics to English verbs; there is also a version adapted to the Spanish language.

represented on the basis of argument structures, that is, structures containing the unit in question and the elements which are part of the semantic frame and which are essential for understanding the meaning of the verb.

In short, *DicoAdventure* combines both approaches, as the *DiCoInfo* and *DiCoEnviro* resources, in order to represent the specialised knowledge of adventure tourism discourse in a comprehensive way. In this chapter, this approach is complemented by the contribution of Morimoto (2001), who analyses the argument structure of Spanish motion verbs on the basis of the conceptual semantics conceived by Jackendoff (1990, 1992). To this end, this chapter is structured as follows. Section 2 explains the fundamental concepts related to argument structure, as well as the main ideas of Mel'čuk and Frame Semantics concerning the definition of the *actant* or *frame element*, that is, the obligatory element without which the definition or the frame evoked would be incomplete. Section 3 describes the cognitive frame used in the *DicoAdventure* resource, which enables the establishment of a set of elements that are usually evoked through verbs and their corresponding argument structures; the type of verbs that are characteristic of the adventure tourism domain are also mentioned. After these sections, which are mainly of a theoretical nature, Section 4 shows the actual analysis of the verbs that frequently appear in adventure tourism texts. The analysis seeks to determine syntactic-semantic patterns based on the discrimination of the arguments presented by these verbs. Finally, Section 5 provides some concluding remarks and future lines of research.

2. The Argument Structure of Verbs

The concept of *argument structure* stems from an earlier idea, that of *syntactic valence*, coined by Lucien Tesnière in his work *Éléments de syntaxe structurale* (1959). This author makes the distinction (not always easy) between obligatory verbal complements (actants) and optional verbal complements (circumstantials), which will be referred to as *arguments* and *adjuncts* here, respectively. As a result of this new syntactic theory, a particular lexicographical resource was born, the valency dictionary, whose most important component is precisely the argument structure item (Domínguez Vázquez, 2018), an element that usually takes the form of a neutralised example or matrix, that is, an example formed by the minimum syntactic structure of a verb expressed through a sentence with no other elements than the syntactically obligatory ones, which are shown by means of general concepts (Rey-Debove, 1971, pp. 303–306).

Tesnière's theory developed significantly in the following decades, especially in Eastern Europe, through authors of the Moscow Semantic School, such as Aleksandr K. Žolkovskij (Žolkovskij & Mel'čuk, 1965), or of the Leipzig School, such as Gerd Wotjak (2006, 2012). Special attention should be paid to the model developed by Igor A. Mel'čuk within the framework of Explanatory Combinatorial Lexicology, whose approaches are rooted in the theories of the Moscow Semantic School and are, in turn, embedded in the so-called Meaning-Text Theory. As a result of this model, Mel'čuk (1984–1999) developed a work with unique characteristics: the *Dictionnaire explicatif et combinatoire du français contemporain*, in which verbs are defined and structured by means of complete sentences or propositions. The argument structure proposed by this author includes semantic arguments, that is, those variable complements that must appear in the definition of the verb to complete its meaning. The argument structure and the definition are thus integrated items: they support each other and cannot be understood without one another. Semantic decomposition makes it possible to recognise those complements of the predicate which are obligatory. Some are lexicalised, forming a fixed part of the meaning, while other components are formed by *slots* which would be occupied by the different variables and which Mel'čuk represents using capital letters. These variables would form the arguments of the semantic argument structure, as shown below with the verb *to lie*:

X lies to Z about Y   X communicates to Z a piece of false information α about Y and X knows that α is false (Mel'čuk & Milićević, 2020, p. 118).

As can be seen, the argument structure precedes the *definiens*, that is, the definition. This is due to the fact that, for Explanatory Combinatorial Lexicology, the *definiendum* (i.e., the term to be defined) is not really the lemma or the canonical form of the infinitive verb but the argument structure itself (Mel'čuk & Milićević, 2020, p. 151; Mel'čuk & Polguère, 2018, pp. 419–420). In other words, there is a semantic equivalence between the argument structure and the definition insofar as the first can be used instead of the second without losing any information.

Based on an article by Mel'čuk published in 2004, L'Homme (2020, pp. 125–126) proposes two rules for distinguishing between arguments and adjuncts:

1. Arguments are necessary to account for the meaning of a predicative lexical unit (henceforth, LU). If, in a situation denoted by an LU, an argumentative participant is removed, the LU cannot express that situation. For example, the argument structure of the verb *print* must include an argument that plays the

role of INSTRUMENT because mention of the INSTRUMENT is necessary to be able to define that verb: x *prints* y *with* z. *A user* x *marks on paper a set of data* y *with the help of a printer* z.[5]

2. Arguments are expressible in the discourse. Arguments often manifest themselves as syntagms syntactically linked to predicative units. The fact that it is not possible to express it syntactically invalidates it as an argument. For example, although the verb *walk* necessarily implies the semantic component of MANNER 'on foot', it cannot become an argument because the sentence **I walked on foot* is ungrammatical. However, the notion of *expressible* should not be confused with that of *expressed*, since arguments can often be implicit in discourse.

While argument structures reflect properties that predicative LUs have, frames or scenarios of Frame Semantics represent patterns that are repeated and evoked by verbs as well as by other grammatical categories, such as nouns or adjectives. The concepts that are part of the frame are called *frame elements* (L'Homme, 2020, p. 45; Ruppenhofer et al., 2016, p. 8). As in Explanatory Combinatorial Lexicology, also in the linguistic representation of cognitive frames we find components that are obligatory and others that are optional. The former are called *core frame elements* and the latter, *peripheral* and *extra-thematic frame elements* (Ruppenhofer et al., 2016, pp. 23–24). In Mel'čuk's theoretical framework, obligatory complements are represented by variables (x, y, z, etc.). However, in FrameNet this is done by indicating their semantic roles, such as AGENT, PATIENT, DIRECTION, DESTINATION, SOURCE, INSTRUMENT, MANNER, among others, inherited from the case theory developed by Fillmore (1968). Therefore, the complements indicated in the structure of the cognitive framework are exclusively semantic in nature.

In contrast to Mel'čuk's approach, the arguments indicated in the argument structure of the resources developed at the Observatoire de linguistique Sens-Texte (OLST) at the University of Montreal, including *DicoAdventure*, are not represented by variables or letters but by semantic roles, in a similar way as in FrameNet. In addition, each semantic argument displays a prototypical term that plays an important role in the cognitive framework through a pop-up window. For example, in *DicoAdventure*, the semantic role PATH present in the argument structure of the verb $cross_1$ (*A* TOURIST *goes across a river, a track, etc.* [PATH]

---

5   The example is taken from *DiCoInfo* ($print_{1b}$) and is also mentioned in L'Homme and San Martín (2016, p. 171).

from one side [Source] to the other [Destination]) is usually represented by the concept *river*.

3. The Cognitive Frame in *DicoAdventure*

The ideal frame that characterises any activity related to adventure tourism comprises a set of idiosyncratic concepts of that subdomain, such as the Agent, generally referring to a Tourist or the person Responsible for an activity; the Action, represented by a verb, often of motion; the Activity itself, the Instrument (or gear) employed in that specific activity, and the Location (Durán-Muñoz, 2016, pp. 234–241). Figure 1 shows a representation of such a frame, as well as the relationships that are established between the different elements.

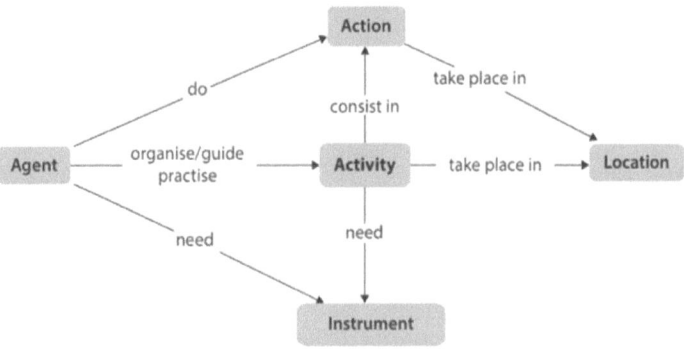

Figure 1. Representation of the prototypical cognitive frame of the adventure tourism domain (Durán-Muñoz, 2016, p. 235).

Durán-Muñoz and L'Homme (2020, p. 46) indicate new subcategories to complete this scenario, specifically linked to the categories of Instrument and Location, in order to account for certain semantic restrictions of the verbs used in the field of adventure tourism. For the former, they specify the existence of three types of vehicles: (1) with an engine (Vehicle_with_Engine), (2) without an engine (Vehicle_without_Engine), and (3) another type involving an animal (Animal); also, for Instrument they specify the frequent use of safety equipment (Safety_Instrument) and different kinds of specific clothing (Clothing). Regarding Location, they distinguish the following subcategories, taking FrameNet as a model: Path, Place, Destination, and Source. It should be noted that the activation of these roles is not always mandatory, especially when the specific activity does not evoke these components.[6]

The verb is the grammatical category in charge of articulating the semantic roles just mentioned, reflecting the frame or setting of the activity through its own argument structure and, therefore, its definition. In this chapter, we propose the existence of a set of argument structures that are shared by all the motion verbs included in the same semantic category. For this purpose, we first propose a classification of the motion verbs extracted from ADVENCOR, a corpus of promotional texts related to the field of adventure tourism,[7] and lemmatised in *DicoAdventure*. Then, we perform an analysis of their semantic structure and their participants.

3.1. Motion verbs in the Spanish adventure tourism domain

Although motion verbs can be studied by considering the semantic frameworks in which they are embedded, in this chapter we rely on the semantic analysis applied by Morimoto (2001) to explain the argument selection criteria presented by the motion verbs lexicalised in the Spanish version of *DicoAdventure* at the time of this study. This author draws several conclusions among which we highlight the following: while inherently directed motion verbs (henceforth, IDMV) include in their lexico-semantic structure the TRAJECTORY component, which can incorporate, in turn, the semantic roles SOURCE and DESTINATION (both telic and atelic), manner-of-motion verbs (henceforth, MMV), on the other hand, are, for the most part, incompatible with those semantic complements which express DESTINATION with a telic notion, that is, those which are introduced by the preposition *a* ['to'], although they can be combined with atelic complements of DESTINATION, which are those expressed by the prepositions *hacia* ['towards'] and *hasta* ['to']. These complements must be interpreted as adjuncts and not as arguments (Morimoto, 2001, p. 128).

> [...] los VVDD ['inherently directed motion verbs'] expresan un desplazamiento con una determinada orientación o dirección, mientras que los VMMS ['manner-of-motion verbs'] como *caminar, correr, nadar*, etc. se limitan a señalar la existencia de

---

[6] One of the most problematic aspects of combining Explanatory Combinatorial Lexicology and Frame Semantics is that the latter presents patterns that are more abstract and detached from the intensional meaning of LUs, while the former is concerned with the syntactic-semantic properties of concrete units.

[7] For more information about the ADVENCOR corpus, see Durán-Muñoz and Jiménez-Navarro (2021).

un desplazamiento, sin concretar, a nivel léxico, qué tipo de trayectoria está implicada en dicho desplazamiento. La afirmación anterior no significa, naturalmente, que no se pueda, por ejemplo, caminar o correr en determinada orientación (cf. *caminar hacia la estación*), sino que, simplemente, el significado léxico de estos verbos no contiene ninguna información acerca de la trayectoria del desplazamiento denotado por ellos. (p. 46)[8]

In order to describe the notion of TRAJECTORY, Morimoto (2001) builds her proposal upon the study about Spanish prepositions undertaken by Morera (1988). In this work, the author distinguishes between the telic trajectory complement, which indicates the trajectory of the destination itself, and the atelic trajectory complement, expressed by means of the prepositions *hasta* ['to'] or *hacia* ['towards'], which indicate the notion of final limit trajectory and of direction, respectively. As already mentioned, this distinction is of great importance because IDMV can express the role DESTINATION by means of the three prepositions previously stated, whereas MMV in Spanish have to rely either on a verb of displacement, as in *ir nadando a la orilla* ['go swimming to the beach'[9]], or indicate the DESTINATION by means of the preposition *hasta* ['to'] or *hacia* ['towards'], as in *nadar hasta la orilla* ['swim to the beach']. In the latter case, the TRAJECTORY component would no longer form part of the semantic structure but would be an adjunct component.

This phenomenon has a strong influence on the methodology proposed in this chapter, both in terms of the definition of the Spanish motion verbs included in *DicoAdventure* and in the expression of their argument structure, since the semantic components present in their semantic structure must also be reflected in it. Based on Morimoto's (2001, p. 64) distinction of two types of events, we propose to define the IDMV by means of the hypernym *desplazarse* ['travel'] (whose argument structure includes the role TRAJECTORY) and the MMV by means of the verb *moverse* ['move'], which does not specify the TRAJECTORY in its argument structure.

For this work, we analysed the 58 motion verbs (cf. Table 1) that were lemmatised in *DicoAdventure* at the time of this study and that show a real displacement of the agent. They all evoke the prototypical cognitive scenario of the adventure tourism domain, explained in Section 3.

---

8   Besides Morimoto (2001), other authors have also recognised the existence of two types of motion verbs, such as Tesnière (1959), Talmy (1975, 1985), and Lamiroy (1991).
9   The back translations provided are as literal as possible to allow readers to understand the lexical, syntactic, and semantic choices in the Spanish examples.

Table 1. Spanish motion verbs lemmatised in *DicoAdventure*

<u>acceder</u>	<u>conducir</u>	<u>llegar</u>	<u>salvar</u>
<u>acercarse</u>	<u>coronar</u>	montar	<u>sobrevolar</u>
<u>adentrarse</u>	<u>cruzar</u>	**nadar**	<u>sortear</u>
<u>alcanzar</u>	<u>descender</u>	navegar	<u>subir</u>
<u>alejarse</u>	<u>desembarcar</u>	palear	**sumergirse**
andar	<u>deslizarse</u>	<u>pasar</u>	**surcar**
<u>ascender</u>	<u>despegar</u>	**pasear**	**surfear**
<u>atravesar</u>	<u>desplazarse</u>	pedalear	<u>trasladarse</u>
<u>avanzar</u>	**destrepar**	**rapelar**	**trepar**
<u>bajar</u>	<u>dirigirse</u>	recorrer	<u>viajar</u>
<u>bordear</u>	<u>entrar</u>	<u>regresar</u>	<u>visitar</u>
bucear	<u>escalar</u>	remar	volar
cabalgar	<u>escapar</u>	remontar	<u>volver</u>
<u>caer</u>	**esquiar**	<u>salir</u>	
caminar	<u>lanzarse</u>	<u>saltar</u>	

Out of these 58 verbs, 35 were classified as IDMV (underlined in Table 1), while 15 verbs were identified as MMV (in bold in Table 1). In addition, we included eight verbs which are not studied by Morimoto (2001) and which do not directly express displacement, but rather an action that causes displacement (the remaining verbs in Table 1) due to their relevance in adventure tourism. Finally, we discarded those verbs depicting fictive motion, which have already been studied by Durán-Muñoz and Jiménez-Navarro (2023) and Jiménez-Navarro and Durán-Muñoz (2024).

4. Analysis of the Lemmatised Verbs in *DicoAdventure*

In the following sections, we describe the argument structures of the selected motion verbs according to their classification into the three groups previously described. Firstly, we analyse IDMV, which are of three types according to the kind of TRAJECTORY expressed. Secondly, we study MMV, which lack the notion of TRAJECTORY in their semantic structure and whose arguments can only be established by means of their semantic decomposition. Finally, we deal with action verbs whose semantic arguments are related to the notions of the VEHICLE or the INSTRUMENT used to cause motion.

4.1. Inherently directed motion verbs (IDMV)

The vast majority of IDMV belongs to the general language. If they are included in a terminological resource such as *DicoAdventure*, it is because they are linked to terms specific to the domain of adventure tourism and because they evoke the same semantic frame.[10] As mentioned in Section 3.1., a total of 35 verbs from our list represent displacement, and almost all of them can be defined through the verb *desplazarse* ['travel'] accompanied by a certain complement of TRAJECTORY. In fact, they can be classified into three main groups in terms of the type of TRAJECTORY described by Morimoto (2001, p. 82):

1. Trajectory expressing SOURCE and/or DESTINATION, either with a preposition (*de* ['from'] for the role SOURCE and *a* ['to'] for the role DESTINATION) or with a direct complement (to express the DESTINATION role). These can be further divided into two subtypes: (1) without boundary crossing (e.g., *acercarse* ['approach'], *alcanzar* ['reach'], *coronar* ['reach the top of'], *despegar* ['take off'], *dirigirse* ['head'], *llegar* ['arrive'], etc.), and (2) with boundary crossing (*acceder* ['enter'], *adentrarse* ['enter'], *entrar* ['enter'], *escapar* ['escape'], *salir* ['leave']).
2. Trajectory with expression of DIRECTION, either with a preposition (*hacia* ['towards'] or *hasta* ['to'] for the role DESTINATION) or with a direct complement (for the role PATH) (*ascender* ['ascend'], *avanzar* ['move forwards'], *bajar* ['go down'], *descender* ['descend'], *remontar* ['go up'], *subir* ['go up']).
3. Transit trajectory (direct complement to express the role PATH) (*atravesar* ['cross'], *bordear* ['go round'], *cruzar* ['cross'], *pasar* ['pass'], *recorrer* ['tour'], *salvar* ['pass'], *sortear* ['pass']).

Next, we analyse the argument structure of the 35 verbs of displacement collected in *DicoAdventure* according to this threefold categorisation.

---

10 This criterion is also used by L'Homme (1998, p. 72) in resources such as *DiCoInfo* or *DiCoEnviro*.

4.1.1. Motion verbs expressing SOURCE and DESTINATION without overcoming a boundary

Prototypical motion verbs are those that can be accompanied by three arguments: AGENT, SOURCE, and DESTINATION. In general terms, a verb whose semantic structure exhibits the semantic component of DESTINATION may also include an empty slot for the notion of SOURCE, although it is normal for each verb to focus on one of the two roles and not on both simultaneously.

4.1.1.1. DESTINATION-focused inherently directed motion verbs

These verbs include and focus their argument structure on the semantic role DESTINATION, although they can also express a SOURCE. The verbs in Table 2 correspond to this category.

Table 2. DESTINATION-focused inherently directed motion verbs

acercarse	desembarcar	llegar	viajar
alcanzar	desplazarse	regresar	visitar
coronar	dirigirse	trasladarse	volver

Three of the verbs in Table 2 display a slightly different semantic structure: *llegar* ['arrive'], *alcanzar* ['reach'], and *coronar* ['reach the top of'], since they express a finished action as opposed to the others, which denote a displacement in process. This aspectual feature affects their combinatory properties, as they cannot join durative temporal expressions (**Coronó la cima durante cinco horas* [**'He/She reached the top for five hours']), and also their definition, as the hypernym *desplazarse* ['travel'] cannot comply with the principle of semantic equivalence here, given that it is a verb that expresses a displacement in process. The verb *llegar* ['arrive'], whose argument structure[11] could be expressed as *Un* TURISTA **llega** *a/hasta un* DESTINO *desde un* ORIGEN ['A TOURIST arrives at/in

---

11 Given its highly condensed nature, in this chapter we adapt the argument structure of DicoAdventure by making some modifications: (1) we replace the substitution symbol ~ with the verb itself in boldface, which is the one expressed by the lemma; (2) we also include articles; (3) semantic arguments are indicated in small caps and, when the semantic role has some kind of semantic restriction, the concept expressing such restriction is indicated in brackets; (4) prepositions governing verbs are shown in small caps; and (5) we show the DIRECTION component in square brackets when it is

a DESTINATION *from a* SOURCE'], would thus function as a hypernym for the definition of the verbs *alcanzar* ['reach'] and *coronar* ['reach the top of']. The last two verbs are transitive, that is, they express the DESTINATION or the locative argument in the form of a nominal phrase (NP) with the function of a direct object (DO). From a semantic viewpoint, they differ in the fact that *coronar* ['reach the top of'] semantically selects places constituted by mountains and, more specifically, by their peaks, while *alcanzar* ['reach'] imposes no semantic restriction in this sense. Their argument structures are shown in (1) and (2):

(1) *coronar* ['reach the top of']: *Un* TURISTA **corona** *un* DESTINO *(una cima o una montaña).* [*'A* TOURIST **reaches the top of** *a* DESTINATION *(a peak or a mountain)'*].

(2) *alcanzar* ['reach']: *Un* TURISTA **alcanza** *un* DESTINO. [*'A* TOURIST **reaches** *a* DESTINATION'].

As for the verb *desembarcar* ['disembark'], it activates three semantic participants within the cognitive scenario: (1) an agent (TOURIST), (2) the place from where the disembarkation takes place (SOURCE), which is usually a vessel, and (3) the place at/in/on which the tourist disembarks (DESTINATION). However, the SOURCE is usually implied:

(3) *desembarcar* ['disembark']: *Un* TURISTA **desembarca** *en un* DESTINO *desde un* ORIGEN *(una embarcación).* [*'A* TOURIST **disembarks** *at/in/on a* DESTINATION *from a* SOURCE *(a vessel)'*].

Other intransitive verbs found in this group are *acercarse* ['approach'], *llegar* ['arrive'], *dirigirse* ['head'], *trasladarse* ['transfer'], and *viajar* ['travel']. They all own two syntactic arguments: the agent that moves (TOURIST) and the locative (DESTINATION), generally expressed by a prepositional phrase (PP). From a semantic perspective, the verb *viajar* ['travel'] is more complex and needs several more arguments than the rest of the verbs focused on the DESTINATION. It is undoubtedly a verb of displacement, but the person who travels does not do so on foot, at least from a prototypical point of view, but in a vehicle. The most complex aspect of this verb is that it can contain several types of locative semantic arguments: (1) when the locative argument is a PP whose linking element is the preposition *por* ['through'], the semantic role it plays is that of a TRAJECTORY composed of one or several points of reference, which can be interpreted as a PATH or a PLACE (e.g., *Esto permite* **viajar por** *caminos sin pavimentar* ['This allows to **travel through** unpaved roads'] [ADVENCOR]); (2) a second role would

---

embedded in the semantics of the verb itself in order to identify the cognitive frame to which it belongs, as it is unlikely that it is expressed at the syntactic level.

be that of DESTINATION, which would be expressed by the preposition *a* or *hasta* ['to'] (e.g., *en Oviedo tendrás opciones para* **viajar hasta** *las sierras cantábricas* ['in Oviedo you will have options to **travel up to** Cantabrian mountain ranges'] [ADVENCOR]); the difference between *viajar a un lugar* ['travel to a place'] and *viajar hasta un lugar* ['travel up to a place'] lies in the fact that the former truly expresses the destination of the journey while the latter is the goal towards which you are travelling, but first passing through other destinations. Like all verbs expressing the role DESTINATION, *acercarse* ['approach'], *llegar* ['arrive'], *dirigirse* ['head'], *trasladarse* ['transfer'], and *viajar* ['travel'] are also compatible with a locative argument expressing SOURCE through the prepositions *de* or *desde* ['from'], although it is usually omitted or, when it is expressed, DESTINATION is omitted as it is implied. The complete argument structure of the verb *viajar* ['travel'] would be as follows:

(4) *viajar* ['travel']: *Un* TURISTA **viaja** *a/hasta un* DESTINO *en un* VEHÍCULO *por un* LUGAR/CAMINO *de/desde un* ORIGEN. ['*A* TOURIST **travels** *to a* DESTINATION *in a* VEHICLE *through a* PLACE/PATH *from a* SOURCE'].

On the other hand, *acercarse* ['approach'], *desplazarse* ['travel'], *dirigirse* ['head'], and *trasladarse* ['transfer'] are also intransitive verbs that use the preposition *a* ['to'] and focus on the DESTINATION, although sometimes there are examples of usage in which the SOURCE is also expressed:

(5) *acercarse* ['approach'], *desplazarse* ['travel'], *dirigirse* ['head'], *trasladarse* ['transfer']: *Un* TURISTA **se acerca/se desplaza/se dirige/se traslada** *a un* DESTINO *desde un* ORIGEN. ['*A* TOURIST **approaches at/travels to/heads for/is transferred to** *a* DESTINATION *from a* SOURCE'].

Finally, *regresar* ['return'] and *volver* ['return'] function identically to the previous verbs, but differ at the semantic level because DESTINATION expresses a place where one has been before:

(6) *regresar* ['return'], *volver* ['return']: *Un* TURISTA **regresa/vuelve** *a un* DESTINO *desde un* ORIGEN. ['*A* TOURIST **returns** *to a* DESTINATION *from a* SOURCE'].

4.1.1.2. SOURCE-*focused inherently directed motion verbs*

Unlike DESTINATION-focused IDMV, the most frequent argument in these verbs is the one that expresses SOURCE, although this does not prevent the argument DESTINATION from also being present. The motion can be performed consciously (e.g., *alejarse* ['move away'], *despegar* ['take off']) or produced by the effect of gravity (e.g., *caer* ['fall'], *lanzarse* ['launch'], *saltar* ['jump']). In *DicoAdventure* we identified five verbs that fall into this category and are listed in Table 3:

Table 3. Source-focused inherently directed motion verbs

| alejarse | caer | despegar | lanzarse | saltar |

With respect to the verbs that appear in Table 3, *alejarse* ['move away'] only presents two semantic arguments, Tourist and Source, and it is a neutral verb without any particular semantic restriction. On the other hand, the verb *despegar* ['take off'], although it includes the same arguments, it evokes other aspects by its own semantic nature: (1) the displacement implies a Direction of an ascending type (although this component is lexicalised and does not appear in the argument structure), in such a way that it separates the Tourist from the ground, and (2) two more arguments are necessary: one that expresses a Vehicle capable of flying and another one that expresses the Destination:

(7) *alejarse* ['move away']: *Un Turista se aleja de un Origen.* ['A Tourist **moves away** from a Source'].
(8) *despegar* ['take off']: *Un Turista despega de un Origen con un Vehículo a/hacia un Destino.* ['A Tourist **takes off** from a Source with a Vehicle to/towards a Destination'].

The rest of the verbs in Table 3, *caer* ['fall'], *lanzarse* ['launch'], and *saltar* ['jump'], are practically synonymous in their specialised meaning. They are polysemous verbs, but in the context of adventure tourism not only do they imply a Source from which to fall, launch (a platform or a plane), or jump, but they also involve a Destination on which to land, as well as the use of a Safety_Instrument, such as harnesses, cables, or parachutes, concepts that would be included in a third semantic argument that would complete the argumentative structure of these verbs:

(9) *caer* ['fall'], *lanzarse* ['launch'], *saltar* ['jump']: *Un Turista cae/se lanza/salta desde un Origen (puente, plataforma, avión) a/hacia un Destino equipado con determinado Instrumento_de_Seguridad (arneses, cables, paracaídas) [en Dirección descendente].* ['A Tourist **falls/launches/jumps** from a Source (bridge, platform, plane) to/towards a Destination equipped with specific Safety_Instrument (harnesses, cables, parachutes) [in downward Direction]'].

It is worth mentioning that the semantic component of Direction in the verbs *caer* ['fall'], *lanzarse* ['launch'], and *saltar* ['jump'] is fully integrated in the meaning of each verb itself. For that reason, as indicated in *DicoAdventure*, it rarely appears in real contexts. Nevertheless, it is indicated because it is an essential part of the meaning of the verbs.

4.1.2. Inherently directed motion verbs expressing DESTINATION with exceeding limit

In this category, we find the five verbs listed in Table 4:

Table 4. Motion verbs expressing DESTINATION with exceeding limit

acceder	adentrarse	entrar	escapar	salir

These verbs can be classified into two different groups. The first group, consisting of the verbs *acceder* ['enter'], *adentrarse* ['enter'], and *entrar* ['enter'], has the notion of TRAJECTORY inwards as its semantic component. Specifically, *acceder* ['enter'] implies moving into a place through a route or doing so in a particular way; *adentrarse* ['enter'] means to move into a place but deeply, without remaining superficial; and, finally, *entrar* ['enter'] is an unmarked verb and, therefore, the most neutral one, as it refers to move into a place. They coincide in the fact that the destinations are usually delimited places, either enclosed, such as caves, or some kind of natural boundary, such as mountain ranges and ravines (e.g., *se puede* **acceder a varias grutas** *submarinas de varios metros* ['we can **enter several underwater caves** of various kilometres']; ***nos adentraremos en la cordillera*** *Cantábrica* ['we will **enter the** Cantabrian **mountain range**']; ***Entraremos en una cueva*** *sencilla* ['we will **enter an** easy **cave**'] [ADVENCOR]). Moreover, the verb *acceder* ['enter'] seems to be frequently combined with a complement of the PATH type, as someone enters a place usually via a certain route. The corpus does not, however, reflect this behaviour with the other two verbs.

(10) *acceder* ['enter']: *Un TURISTA* **accede** *a un DESTINO (un lugar delimitado, como cuevas o grutas, pero también cordilleras y barrancos) a través de un CAMINO (vías ferratas, escaleras, toboganes).* ['*A TOURIST* **enters** *a DESTINATION (a delimited place, such as caves, but also mountain ranges and ravines) through a PATH (vías ferrata, stairs, toboggans)*'].

(11) *adentrarse* ['enter']: *Un TURISTA* **se adentra** *en un DESTINO.* ['*A TOURIST* **enters** *a DESTINATION*'].

(12) *entrar* ['enter']: *Un TURISTA* **entra** *en un DESTINO.* ['*A TOURIST* **enters** *a DESTINATION*'].

The second group of verbs differs from the previous one because the notion of 'overcoming the limit' is from inside out, and is composed of the verbs *escapar* ['escape'] and *salir* ['leave']. In the verb *escapar* ['escape'], a process of metaphorisation can be observed, according to which the SOURCE (e.g., the city, the routine, the bustle, or the winter) is compared to a prison (e.g., *Escápese del bullicio de la ciudad* ['Escape from the bustle of the city']; *Absolutamente recomendable para escapar de la rutina* ['Absolutely recommendable to **escape from the routine**'] [ADVENCOR]), which is not the case of the verb *salir* ['leave']. Their argument structures would, however, be identical:

> (13) *escapar* ['escape'], *salir* ['leave']: *Un TURISTA **escapa/sale** de un ORIGEN (lugar cerrado).* ['A TOURIST **escapes from/leaves** a SOURCE (enclosed place)'].

4.1.3. Inherently directed motion verbs expressing DIRECTION

The third group of IDMV analysed here incorporates the six verbs shown in Table 5, which are characterised by including as a participant the semantic notion of DIRECTION, either vertical or horizontal:

Table 5. Inherently directed motion verbs expressing DIRECTION

ascender	bajar	remontar
avanzar	descender	subir

Verbs expressing DIRECTION of a vertical type, such as *subir* ['go up'] or *bajar* ['go down'], or of a horizontal type, such as *avanzar* ['move forwards'], are potentially transitive, where the NP with the function of DO is played by the place of reference. Regarding DIRECTION, this is, once again, an intrinsic semantic component, as it is fully integrated into the semantics of the verbs[12] and can hardly be expressed in context. The role DESTINATION can be preceded in this type of verbs by the preposition *hacia* ['towards'], as long as the DESTINATION coincides with the actual DIRECTION expressed by the verb:

> (14) *ascender* ['ascend'], *bajar* ['go down'], *descender* ['descend'], *subir* ['go up'], *remontar* ['go up']: *Un TURISTA **asciende/baja/desciende/sube/remonta** un CAMINO/por un CAMINO (vía en pendiente) hacia un DESTINO desde un ORIGEN [en DIRECCIÓN ascendente/descendente].* ['A TOURIST **ascends/goes down/descends/goes up** a PATH/through a PATH (on a slope) towards a DESTINATION from a SOURCE [in upward/downward DIRECTION]'].

---

12 English verbs do not act in this way, as they can express in the surface syntactic structure certain adverbial phrases indicating DIRECTION, such as *to climb up/uphill/down, to abseil down/downhill*.

(15) *avanzar* ['move forwards']: *Un Turista avanza por un Camino hacia un Destino (lugar situado en Dirección hacia adelante) desde un Origen [en Dirección hacia adelante]*. ['*A Tourist moves forwards through a Path towards a Destination (a place located in forward Direction) from a Source [in forward Direction]*'].

4.1.4. Inherently directed motion verbs expressing transit

Unlike verbs expressing Source and Destination with boundary crossing, idmv expressing transit imply that the boundary itself, expressed through an NP with a DO function, is another place to be crossed. Therefore, four arguments are identified, in addition to the Agent: (1) the Place or point of reference which acts as a boundary, (2) the Path (*footbridge* or *bridge*) which crosses that same boundary, (3) the Destination or point of arrival, and (4) the Source, although the last two are so imbricated in the very meaning of the verbs that they often appear only implicitly in the sentence. Moreover, they are incompatible with the expression of telic Destination introduced by the preposition *a* ['to'] (with the partial exception of the verb *cruzar* ['cross'], which does admit examples such as *Cruzó al otro lado* ['He/She crossed to the other side']). The verbs found in this category are seven (cf. Table 6):

Table 6. Inherently directed motion verbs expressing transit

atravesar	cruzar	recorrer	sortear
bordear	pasar	salvar	

The complete structure of idmv expressing transit is expressed as follows:

(16) *atravesar* ['cross'], *bordear* ['go round'], *cruzar* ['cross'], *pasar* ['pass'], *recorrer* ['tour'], *salvar* ['pass'], *sortear* ['pass']: *Un Turista atraviesa/bordea/cruza/pasa/recorre/salva/sortea un Lugar (un límite, como un río o un barranco) por un Camino (pasarela o puente) desde un Origen hasta/hacia un Destino*. ['*A Tourist crosses/goes round/passes/tours a Place (a limit, such as a river or a ravine) through a Path (footbridge or bridge) from a Source to/towards a Destination*'].

4.2. Manner-of-motion verbs (mmv)

Unlike idmv, mmv are more specific to adventure tourism, so they are usually accompanied by concepts related to this domain. The mmv incorporate a semantic component which is Manner to the detriment of the expression Trajectory. This means that, although they denote a displacement in space, the specificity of the trajectory is alien to the semantic structure of the verb. Its definition can be headed by the hypernym *moverse* ['move'] and, in its argument structure, we can find the following arguments: AGENT, PLACE/PATH, VEHICLE, and SAFETY_INSTRUMENT or CLOTHING (these last two would actually be specific semantic subcomponents of the MANNER component).

MMV rarely express a DESTINATION and, when they do, it is through the prepositions *hacia* ['towards'] or *hasta* ['to']. Morimoto (2001, pp. 131–132) does not hesitate to consider such atelic complements as adjuncts. This consideration is reinforced by the fact that this type of TRAJECTORY does not involve any distinctive semantic component and, moreover, is rarely explicit in the syntactic surface structure. The most frequent semantic arguments of MMV are AGENT and PLACE/PATH. The latter connect the action with places evoked within the adventure tourism scenario. In addition, the roles VEHICLE, CLOTHING, and SAFETY_INSTRUMENT have to be added for certain verbs.

MMV can be divided into two main groups according to their surface syntactic structure: transitive MMV and intransitive MMV.

4.2.1. Transitive manner-of-motion verbs

If the semantic component PATH is understood as the extension through which the AGENT moves, there are certain verbs that express this component through an NP with a DO function, although they can also do so through a PP. The transitive MMV found in *DicoAdventure* are listed in Table 7:

Table 7. Transitive manner-of-motion verbs

destrepar	*rapelar*	*surcar*	*trepar*
escalar	*sobrevolar*	*surfear*	

The verbs *escalar* ['climb'] and *trepar* ['climb'] express movement up a vertical slope in an upward direction. In turn, *rapelar* ['abseil'] and *destrepar* ['climb down'] represent a movement in a downward direction. In these pairs of verbs, the former (i.e., *escalar* ['climb'] and *rapelar* ['abseil']) represent an activity that is carried out with a SAFETY_INSTRUMENT, whereas the latter (i.e., *trepar* ['climb'] and *destrepar* ['climb down']) refer to an activity that is done with the help of hands and feet only. The four of them can be either transitive (as in *escalar una montaña* ['climb a mountain']) or intransitive (e.g., *rapelar a través de un cañón* ['abseil through a canyon']):

(17) *escalar* ['climb']: *Un* TURISTA **escala** *un* CAMINO/LUGAR *con la ayuda de un* INSTRUMENTO_DE_SEGURIDAD *[en* DIRECCIÓN *ascendente]*. ['A TOURIST **climbs** a PATH/PLACE *with the help of a* SAFETY_INSTRUMENT *[in upward* DIRECTION*]*'].

(18) *trepar* ['climb']: *Un* TURISTA **trepa** *(por) un* CAMINO/LUGAR *[en* DIRECCIÓN *ascendente]*. ['A TOURIST **climbs** *(through) a* PATH/PLACE *[in upward* DIRECTION*]*'].

(19) *destrepar* ['climb down']: *Un* TURISTA **destrepa** *un* CAMINO/LUGAR *[en* DIRECCIÓN *descendente]*. ['A TOURIST **climbs down** a PATH/PLACE *[in downward* DIRECTION*]*'].

With respect to *rapelar* ['abseil'], this can be understood, in a way, as the antonym of *escalar* ['climb'], since the arguments PATH and DIRECTION coincide, but what changes is the direction, which is downwards in the case of *rapelar* ['abseil']:

(20) *rapelar* ['abseil']: *Un* TURISTA **rapela** *(por) un* CAMINO/LUGAR *con la ayuda de un* INSTRUMENTO_DE_SEGURIDAD *[en* DIRECCIÓN *descendente]*. ['A TOURIST **abseils** *(through) a* PATH/PLACE *with the help of a* SAFETY_INSTRUMENT *[in downward* DIRECTION*]*'].

On the other hand, the MMV *sobrevolar* ['overfly'], *surcar* ['sail'], and *surfear* ['surf'] are verbs that exhibit similar semantic structures: (1) *sobrevolar* ['overfly'] implies a PLACE above which the AGENT moves at a certain height; (2) *surcar* ['sail'] also contains the PLACE role, but this time the displacement takes place directly over it, either a water surface or the air, so it is crossed by splitting it in two, figuratively speaking; and (3) *surfear* ['surf'] is similar to *surcar* ['sail'], but the former imposes a semantic restriction by selecting a PLACE referring to a water surface, more specifically, a wave or waves. Table 8 shows the semantic components of these verbs from the point of view of the distinctive function of their components:

Table 8. Semantic roles of transitive manner-of-motion verbs and their distinctive value

Verb	AGENT	PLACE	MANNER
Destrepar ['climb down']	TOURIST	Mountain	With hands and feet
Escalar ['climb']	TOURIST	Mountain	With SAFETY_INSTRUMENT
Rapelar ['abseil']	TOURIST	Mountain	With SAFETY_INSTRUMENT
Sobrevolar ['overfly']	TOURIST	Landscape	With VEHICLE_WITH_ENGINE / VEHICLE_WITHOUT_ENGINE
Surcar ['sail']	TOURIST	Air environment/ water surface	With VEHICLE_WITHOUT_ENGINE
Surfear ['surf']	TOURIST	Water surface	With VEHICLE_WITHOUT_ENGINE
Trepar ['climb']	TOURIST	Mountain/tree	With hands and feet

4.2.2. Intransitive manner-of-motion verbs

The main difference between transitive (cf. Table 7) and intransitive (cf. Table 9) MMV is the way in which they syntactically express their semantic structure (with or without a DO, respectively). Apart from that, they follow the same scheme at the logical-semantic level: an AGENT moves with an indeterminate trajectory through a PLACE/PATH and in a certain MANNER.

Table 9. Intransitive manner-of-motion verbs

andar	caminar	esquiar	pasear
bucear	deslizarse	nadar	sumergirse

The verbs *andar* ['walk'], *caminar* ['walk'], and *pasear* ['walk'] are practically synonymous and function syntactically the same in Spanish. They could be described as follows: *Un TURISTA **se mueve** a pie a un ritmo tranquilo por un CAMINO/en un LUGAR* ['A TOURIST **moves** on foot at an unhurried pace through a PATH/in a PLACE']. As can be seen, the semantic component MANNER is shown through the adverbial locutions *a pie* ['on foot'] (as opposed to other verbs such as *volar* ['fly'] or *navegar* ['sail']) and *a un ritmo tranquilo* ['at an unhurried pace'] (to differentiate these verbs from others such as *correr* ['run'], not lemmatised in DicoAdventure yet). However, this semantic component is fully integrated into the semantics of the verbs. On the other hand, the complements PATH and PLACE are considered arguments because they establish a connection between the verb and the prototypical places where the adventure tourism activity takes place:

(21) *andar* ['walk'], *caminar* ['walk'], *pasear* ['walk']: *Un TURISTA **anda/camina/pasea** por un CAMINO/en un LUGAR.* ['A TOURIST **walks** through a PATH/in a PLACE'].

As for the verbs *deslizarse* ['slide'] and *esquiar* ['ski'], the former entails a slippery PATH sloping downwards, so the MANNER assumes a type of posture that allows sliding. The verb *esquiar* ['ski'], on the other hand, implies a snowy PATH which requires the use of skis (INSTRUMENT understood as a subcategory of MANNER):

(22) *deslizarse* ['slide']: *Un TURISTA **se desliza** por un CAMINO (vía en pendiente).* ['A TOURIST **slides** through a PATH (on a slope)'].

(23) *esquiar* ['ski']: *Un TURISTA **esquía** con la ayuda de un determinado INSTRUMENTO (esquíes) por un CAMINO/en un LUGAR.* ['A TOURIST **skis** with the help of a certain INSTRUMENT (skis) through a PATH/in a PLACE'].

Finally, the MMV whose activity takes place in an aquatic environment are *bucear* ['dive'], *nadar* ['swim'], and *sumergirse* ['plunge']. The verbs *bucear* ['dive'] and *nadar* ['swim'] are intransitive verbs that express the semantic component PLACE through a PP introduced by the preposition *en* ['into']. The difference between them is that, for the verb *bucear* ['dive'], the PLACE it selects is always located below the surface of the water, whereas *nadar* ['swim'] expresses a displacement on the surface of the water itself. The MANNER is partly identical since both diving and swimming require moving arms and legs using a technique. Nevertheless, *bucear* ['dive'] is an action that is often carried out with the help of certain equipment, such as a snorkel, an oxygen tank, or fins. This information is compulsory because it activates a part of the adventure tourism scenario, namely SAFETY_INSTRUMENT, and, therefore, these indications are necessary to know how the action is usually performed. As for the verb *sumergirse* ['plunge'], like *escalar* ['climb'], it involves a type of displacement that also governs a semantic complement of DESTINATION (into the water):

(24) *bucear* ['dive']: *Un* TURISTA **bucea** *en un* LUGAR *(medio acuático) con determinada* VESTIMENTA *o determinado* INSTRUMENTO_DE_SEGURIDAD *(snorkel, bombona de oxígeno, aletas, etc.).* ['A TOURIST **dives** in a PLACE (aquatic environment) with certain CLOTHING or a certain SAFETY_INSTRUMENT (snorkel, oxygen tank, fins, etc.)'].

(25) *nadar* ['swim']: *Un* TURISTA **nada** *en un* LUGAR *(superficie acuática).* ['A TOURIST **swims** in a PLACE (aquatic surface)'].

(26) *sumergirse* ['plunge']: *Un* TURISTA *se* **sumerge** *en un* DESTINO *(medio acuático).* [A TOURIST **plunges** into a DESTINATION (aquatic environment)'].

4.3. Action verbs (AV)

In order to complete the description of all the verbs lemmatised in *DicoAdventure*, we include here a third type of motion verbs that is not studied by Morimoto (2001) because they are not strictly speaking motion verbs. They are action verbs that involve travelling in a vehicle. Unlike other verbs previously analysed that include the semantic role VEHICLE in their argument structure, such as *viajar* ['travel'], *volar* ['fly'], or *surfear* ['surf'], these verbs, shown in Table 10, imply an actual control of the vehicles used when moving.

Table 10. Action verbs

cabalgar	montar	palear	remar
conducir	navegar	pedalear	volar

From a semantic point of view, there are two variants: in the first one, the verbs express direct control of a vehicle, such as *cabalgar* ['ride an ANIMAL']), *conducir* ['drive'], *navegar* ['sail'], *montar* ['ride'], and *volar* ['fly']; in the second type, the verb expresses the use of one or more instruments which drive the vehicle (*palear* ['paddle'], *pedalear* ['pedal'], and *remar* ['row']). Furthermore, from a syntactic perspective, these eight verbs can also be classified into transitive and intransitive verbs, as reflected in Sections 4.3.1. and 4.3.2., respectively.

4.3.1. Transitive action verbs

The transitive AV are *cabalgar* ['ride an ANIMAL'] and *conducir* ['drive']. These are characterised by the selection of the vehicle in which the TOURIST rides through an NP with the function of a DO. The role VEHICLE is argumentative not only in terms of the syntactic function it plays but also because it is a distinctive semantic component; in other words, *cabalgar* ['ride an ANIMAL'] is restricted to animals of the equine family, while *conducir* ['drive'] semantically selects vehicles that travel on land, excluding any kind of animal. As part of the prototypical cognitive framework of the adventure tourism domain, a third semantic (but not syntactic) argument can be included, relating to PATH if it is a PP introduced by the preposition *por* ['through'] (e.g., *También puedes **cabalgar por el valle de Jalcomulco*** ['You can also **ride through the valley of Jalcomulco**'] [ADVENCOR]) or relating to PLACE if the PP has the preposition *en* ['in'] as a connecting element (e.g., *Aún recuerdo la sensación de **conducir** la moto de nieve **en la zona ártica*** ['I still remember the feeling of **driving** the snowmobile **in the arctic zone**'] [ADVENCOR]). Therefore, their argumentative structures look like this:

(27) *cabalgar* ['ride an ANIMAL']: *Un TURISTA **cabalga** un ANIMAL (caballo) por un CAMINO o en un LUGAR.* ['A TOURIST **rides** an ANIMAL (horse) through a PATH or in a PLACE'].

(28) *conducir* ['drive']: *Un TURISTA **conduce** un VEHÍCULO (terrestre) por un CAMINO o en un LUGAR.* ['A TOURIST **drives** a VEHICLE (on land) through a PATH or in a PLACE'].

4.3.2. Intransitive action verbs

The intransitive AV are *montar(se)* ['get on/in'], *navegar* ['sail'], *palear* ['paddle'], *pedalear* ['pedal'], *remar* ['row'], and *volar* ['fly']. As in the previous case, these verbs can express the control of a vehicle and they almost always do so through a PP headed by the preposition *en* ['in'].

The verb *montar(se)* is polysemous: it can mean 'to get on/in a vehicle', 'to occupy the seat of a vehicle', or 'to control the movements of a certain type of vehicle'. It is the context that determines its semantic value, although it is true that when the vehicle has room for only one passenger, such as a bicycle, a horse, or sometimes a kayak, then the meaning of the verb *montar(se)* is unequivocally 'to control the movements of a certain type of vehicle' ['ride']. The same idea can also be applied to the verbs *navegar* ['sail'] and *volar* ['fly'].

(29) *montar(se)* ['get on/in']: *Un* TURISTA *(se) monta en un* VEHÍCULO *o (en) un* ANIMAL *(caballo) por un* CAMINO *o en un* LUGAR. ['A TOURIST **gets on/in** a VEHICLE or an ANIMAL (horse) through a PATH or in a PLACE'].

(30) *navegar* ['sail']: *Un* TURISTA *navega en un* VEHÍCULO *(vehículo para navegar) en un* LUGAR *(superficie acuática)*. ['A TOURIST **sails** in a VEHICLE (a vehicle to sail) in a PLACE (aquatic surface)'].

(31) *volar* ['fly']: *Un* TURISTA *vuela en un* VEHÍCULO *(vehículo capaz de volar) por un* LUGAR *(medio aéreo)*. ['A TOURIST **flies** in a VEHICLE (a vehicle which can fly) through a PLACE (air environment)'].

The other three verbs, *palear* ['paddle'], *pedalear* ['pedal'], and *remar* ['row'], exhibit the same argument structure as the previous ones, but they show some differences at the semantic level. The main difference is that the superordinate unit of these verbs is not the verb *controlar* ['control'], but the verb *mover* ['move']. The change of hypernym obeys the principle of synonymy applied to the definition: in examples such as *pedalea más rápido* ['pedal quicker'] or *rema más fuerte* ['row more briskly'], it is clear that the verb *controlar* ['control'] cannot replace them, but the hypernym *mover* ['move'] can: *mueve los pedales más rápido* ['move the pedals quicker'] or *mueve los remos con más fuerza* ['move the oars more briskly']. *Palear* ['paddle'], *pedalear* ['pedal'], and *remar* ['row'] have in common that they select the VEHICLE through a PP headed by the preposition *en* ['in'] (e.g., **Palearemos en kayak** [**we will paddle a kayak**']; **Pedaleando en bicicleta** ['**Pedalling a bicycle**'] [ADVENCOR]). Semantically, the VEHICLE referred to by the verb *pedalear* ['pedal'] is a bicycle, while *palear* ['paddle'] and *remar* ['row'] would select a type of craft. More important, however, would be the argument MANNER, namely, INSTRUMENT. While *pedalear* ['pedal'] contains the fully lexicalised complement INSTRUMENT in such a way that it cannot be expressed in the argument structure (one does not usually say **alguien pedalea con pedales* [*'somebody pedals with pedals']), the instruments in the case of *palear* ['paddle'] and *remar* ['row'] can be made explicit, especially when they are expressed by means of modifiers (e.g., *palear con una pala de hoja única* ['paddling with a single blade paddle'], *palear con una pala de hoja doble* ['paddling with a double-bladed paddle'], *remar con un remo de cuchara simple* ['rowing with a spoon-bladed oar'], *remar con una pala de doble cuchara* ['rowing with a double-bladed oar']):

(32) *palear* ['paddle']: *Un* TURISTA **palea** *en un* VEHÍCULO_SIN_MOTOR *(kayak) con un* INSTRUMENTO *(pala) por un* CAMINO *(superficie acuática).* ['A TOURIST **paddles** a VEHICLE_WITHOUT_ENGINE *(kayak) with an* INSTRUMENT *(paddle) through a* PATH *(aquatic surface)*'].

(33) *pedalear* ['pedal']: *Un* TURISTA **pedalea** *en un* VEHÍCULO_SIN_MOTOR *(bicicleta) por un* CAMINO. ['A TOURIST **pedals** a VEHICLE_WITHOUT_ENGINE *(bicycle) through a* PATH'].

(34) *remar* ['row']: *Un* TURISTA **rema** *en un* VEHÍCULO_SIN_MOTOR *(kayak) con un* INSTRUMENTO *(uno o varios remos) por un* CAMINO *(superficie acuática).* ['A TOURIST **rows** *in a* VEHICLE_WITHOUT_ENGINE *(kayak) with an* INSTRUMENT *(one or several oars) through a* PATH *(aquatic surface)*'].

5. Conclusions

In this chapter, we aimed to study the argument structures of a selected group of motion verbs in the field of adventure tourism and classify these units in terms of Morimoto's (2001) proposal. We were also inspired by Mel'čuk's (Mel'čuk, 2004; Mel'čuk & Polguère, 2018; Mel'čuk et al., 1995) work in the context of Explanatory Combinatorial Lexicology and by some principles of Frame Semantics (Fillmore, 1977, 1982, 1985), the two main theoretical perspectives which are the basis of *DicoAdventure*.

Throughout this work we have proved that there are three main groups of motion verbs used in the field of adventure tourism:

1. Inherently directed motion verbs (IDMV), whose argument structure can be summarised as follows: AGENT + verb + TRAJECTORY [SOURCE/DESTINATION/PATH]. This group includes five subgroups:
   a. DESTINATION-focused verbs: *acercarse, alcanzar, coronar, de*s*embarcar, desplazarse, dirigirse, llegar, regresar, trasladarse, viajar, visitar, volver.*
   b. SOURCE-focused verbs: *alejarse, caer, despegar, lanzarse, saltar.*
   c. Verbs of destination with exceeding limit: *acceder, adentrarse, entrar, escapar, salir.*
   d. Directional verbs: *ascender, avanzar, bajar, descender, remontar, subir.*
   e. Verbs expressing transit: *atravesar, bordear, cruzar, pasar, recorrer, salvar, sortear.*

2. Manner-of-motion verbs (MMV), which offer a semantic structure in which the role TRAJECTORY is not well defined and, on the contrary, the MANNER is semantically included in the meaning of the verb. Moreover, the roles PLACE and PATH are included given their relevance in the field of adventure tourism. Its semantic structure can be expressed in the following way: AGENT + verb

+ Manner [Instrument/Vehicle/Clothing/Safety_Instrument] + Place/Path. According to their syntactic behaviour, these verbs can be divided into two types:
  a. Transitive verbs: *destrepar, escalar, rapelar, sobrevolar, surcar, surfear, trepar.*
  b. Intransitive verbs: *andar, caminar, bucear, deslizarse, esquiar, nadar, pasear, sumergirse.*

3. Action verbs: this third type of verbs are not, strictly speaking, motion verbs, but they express an action that produces the agent's movement. Here we can distinguish two types:
   a. Verbs that express the control of a vehicle or an animal. In turn, these can be either transitive (*cabalgar, conducir*) or intransitive (*montar(se), navegar, volar*).
   b. Verbs that express the displacement of an instrument that is required to be moved, like *palear, pedalear,* and *remar*. All of them are intransitive.

For future work, analyses might focus on the role of adjuncts in the formation of both Spanish verb collocations and complex structures that help *DicoAdventure* users to acquire specialised knowledge. Similarly, there is a need for work that explores, from a contrastive point of view, the argument structures of equivalent verbs in English, as the syntactic and semantic behaviour often varies from one language to another, especially when it comes to verbs that express the way of moving. This information can be particularly useful for users, especially translators, when consulting *DicoAdventure*.

References

Domínguez Vázquez, M. J. (2018). Was sind Valenzwörterbücher? *Sprachwissenschaft, 43*(3), 309–342.

Durán-Muñoz, I. (2016). Producing frame-based definitions: A case study. *Terminology, 22*(2), 224–250. https://doi.org/10.1075/term.22.2.04mun

Durán-Muñoz, I. (2021). *DicoAdventure* y la terminología del turismo de aventura: Propuesta de diccionario en línea. In T. Barceló Martínez, I. Delgado Pugés, & F. García Luque (Eds.), *Tendencias actuales en traducción especializada, traducción audiovisual y accesibilidad* (pp. 395–417). Tirant Lo Blanch.

Durán-Muñoz, I., & Jiménez-Navarro, E. L. (2021). Colocaciones verbales en el turismo de aventura: Estudio contrastivo inglés-español. In G. Corpas Pastor, M.ª R. Bautista Zambrana, & C. M. Hidalgo-Ternero (Eds.), *Sistemas *fraseológicos en contraste: Enfoques computacionales y de corpus* (pp. 121–142). Comares.

Durán-Muñoz, I., & Jiménez-Navarro, E. L. (2023). Motion verbs in adventure tourism: A lexico-semantic approach to fictive meaning. *IJES, 23*(1), 27–48. https://doi.org/10.6018/ijes.532851

Durán-Muñoz, I., & L'Homme, M.-C. (2020). Diving into English motion verbs from a lexico-semantic approach: A corpus-based analysis of adventure tourism. *Terminology, 26*(1), 33–59. https://doi.org/10.1075/term.00041.dur

Fillmore, C. (1968). The case for case. In E. W. Bach & R. T. Harms (Eds.), *Universals in linguistic theory* (pp. 1–25). Holt, Rinehart & Winston.

Fillmore, C. J. (1977). Scenes and Frames Semantics. In A. Zampolli (Ed.), *Linguistic structures processing* (pp. 55–88). North Holland Publishing.

Fillmore, C. J. (1982). Frame Semantics. In Linguistic Society of Korea (Ed.), *Linguistics in the morning calm* (pp. 111–137). Hanshin Publishing Company.

Fillmore, C. J. (1985). Frames and the semantics of understanding. *Quaderni di Semantica, 6*(2), 222–254.

Jackendoff, R. (1990). *Semantic structures*. MIT Press.

Jackendoff, R. (1992). *Languages of the mind. Essays on mental representation*. MIT Press.

Jiménez-Navarro, E. L. (2020). *Treatment and representation of verb collocations in the specialised language of adventure tourism* [Doctoral dissertation, Universidad de Córdoba]. Helvia. https://helvia.uco.es/xmlui/handle/10396/20976

Jiménez-Navarro, E. L., & Durán-Muñoz, I. (2024). Collocations of fictive motion verbs in adventure tourism: A corpus-based study of the English language. *Revista Española de Lingüística Aplicada* (online). https://doi.org/10.1075/resla.21042.jim

Lamiroy, B. (1991). *Léxico y gramática del español. Estructuras verbales de espacio y de tiempo*. Anthoropos.

L'Homme, M.-C. (1998). Le statut du verbe dans le texte specialisé. *Cahiers de Lexicologie, 73*(2), 61–84.

L'Homme, M.-C. (2007). Using explanatory and combinational lexicology to describe terms. In L. Wanner (Ed.), *Selected lexical and grammatical topics in the Meaning-Text Theory. In Honour of Igor Mel'cuk* (pp. 11–50). John Benjamins Publishing Company. https://doi.org/10.1075/tlrp.20

L'Homme, M.-C. (2020). *Lexical semantics for terminology. An introduction*. John Benjamins Publishing Company. https://doi.org/10.1075/tlrp.20

L'Homme, M.-C., & San Martín, A. (2016). Définition terminologique: Systématisation de règles de rédaction dans les domaines de l'informatique et de l'environnement. *Cahiers de lexicologie, 109*(2), 145–172. https://doi.org/10.15122/isbn.978-2-406-06861-7.p.0145

Mel'čuk, I. (Dir.) (1984–1999). *Dictionnaire explicatif et combinatoire du français contemporain. Recherches lexico-sémantiques* [Volume I (1984); Volume II (1988); Volume III (1992); Volume III (1999)]. Presses de l'Université de Montréal.

Mel'čuk, I. (2004). Actants in semantics and syntax I: Actants in semantics. *Linguistics, 42*(1), 1–66. https://doi.org/10.1515/ling.2004.004

Mel'čuk, I., Clas, A., & Polguère, A. (1995). *Introduction à la lexicologie explicative et combinatoire*. Duculot.

Mel'čuk, I., & Milićević, J. (2020). *An advanced introduction to semantics. A meaning-text approach*. Cambridge University Press.

Mel'čuk, I., & Polguère, A. (2018). Theory and practice of lexicographic definition. *Journal of Cognitive Science, 19*(4), 417–470. https://doi.org/10.17791/jcs.2018.19.4.417

Morera, M. (1988). *Estructura semántica del sistema preposicional del español moderno y sus campos de usos*. Servicio de Publicaciones del Excmo. Cabildo Insular de Fuerteventura.

Morimoto, Y. (2001). *Los verbos de movimiento*. Visor Libros.

Rey-Debove, J. (1971). *Étude linguistique et sémiotique des dictionnaires français contemporains*. Mouton.

Ruppenhofer, J., Ellsworth, M., Petruck, M. R. L., Johnson, Ch. R., Baker, C. F., & Scheffczyk, J. (2016). *FrameNet II: Extended Theory and Practice*. International Computer Science Institute.

Talmy, L. (1975). Semantics and syntax of motion. In J. P. Kimball (Ed.), *Syntax and semantics* (pp. 181–238). Academic Press.

Talmy, L. (1985) Lexicalization patterns: Semantic structure in lexical forms. In T. Shopen (Ed.), *Language typology and syntactic description*, 3: *Grammatical categories and the lexicon* (pp. 57–150). Cambridge University Press.

Tesnière, L. (1959). *Eléments de syntaxe structurale*. Klincksiek.

Wotjak, G. (2006). Semántica léxica y sintaxis: Verbos en la encrucijada entre pragmática y cognición. *Signo y Seña, 15,* 43–74. https://doi.org/10.34096/sys.n15.5814

Wotjak, G. (2012). Valencia y colocabilidad: Aspectos cognitivos-semánticos, morfosintácticos y pragmático-situativos. In T. E. Jiménez Juliá, B. López Meirama, V. Vázquez Rozas, & A. Veiga Rodríguez (Eds.), *Cum corde et in nova grammatica: Estudios ofrecidos a Guillermo Rojo* (pp. 897–927). University of Santiago de Compostela.

Žolkovskij, A. K., & Mel'čuk, I. A. (1965). O vozmožnom metode i instrumentax semantičeskogo sinteza. *Naučno-texničeskaja Informacija*, 5, 23–28.

Acknowledgements

This research work has been carried out within the framework of the R&D project "DicoAdventure: diseño y desarrollo de un recurso electrónico especializado bilingüe (inglés, español) sobre el turismo de aventura a partir de marcos semánticos" (Ref. UCO-1380857-F), co-funded by the Operational Programme FEDER 2014–2020 and the Consejería de Economía, Conocimiento, Empresas y Universidad of the Andalusian regional government.

Eva Lucía Jiménez-Navarro

Universidad de Córdoba
lucia.jimenez@uco.es

6 Prepositional phrase collocations of motion verbs: A corpus-driven study in adventure tourism

Abstract: This chapter aims to contribute to the characterisation of the language used in the domain of adventure tourism. For so doing, it focuses on its phraseological component and explores the prepositional phrases collocating with a set of motion verbs extracted from a specialised corpus called ADVENCOR (cf. Durán-Muñoz & Jiménez-Navarro, 2021). The methodology applied is corpus-driven and follows three main steps: (1) extraction of motion verbs, (2) extraction of prepositional phrase collocations, and (3) classification of the collocations selected according to the type of motion (i.e., real or fictive) described. The findings reveal that collocations showing real motion were more common than those depicting fictive motion. Albeit some verbs produced (even the same) collocations representing both types of events, prepositions were used to enhance the meaning implied by the verbs in real motion cases rather than in fictive ones. Besides, an analysis of the semantic roles expressed by the prepositions was undertaken, which demonstrated that arguments of the verbs were more common to appear in collocations expressing fictive motion, whereas those related to real motion featured both arguments and circumstantials. Finally, the most recurrent preposition was the same in both cases, *through*, which introduces the PATH where the motion occurs.

Keywords: adventure tourism, collocation, corpus-driven study, motion verb, prepositional phrase

1. Introduction

It could be stated that an array of studies conducted in the 1940s marked a significant turning point in the field of phraseology. At this time, a group of Russian scholars led by V. V. Vinogradov (1947) purported to classify multi-word combinations according to two main criteria, their degree of fixedness and their idiomaticity. Basically, these word groupings were considered to be located along a continuum where unmotivated fixed units were found at one end (e.g., *once in a blue moon, to be on the carpet*) and motivated flexible units at the other (e.g., *to eat fish, to cook rice*). In the middle of this series, collocations gained the status of two-word combinations which were partially semantically opaque and partially structurally flexible (e.g., *to meet the demand, to make a mistake*), and for this reason they have sparked the interests of researchers in the last decades.

As it is explained in Section 2, there is a wide variety of definitions concerning the term *collocation*; despite that, several features can be identified when it comes to their inclusion in a framework of phraseological units (henceforth, PUs). Additionally, irrespective of whether they are tackled from a general perspective or from a specialised perspective, their concept remains the same. To put it simple, they are combinations of two words which are conferred a different status, since one of them is usually chosen first and evokes the use of the second word. Normally, collocations containing nouns have attracted the most attention, probably because nouns are considered the most productive category in terminology (Vargas Sierra, 2012). However, this work is centred on verb collocations, given that previous research on the segment of adventure tourism has shown that motion verbs play a pivotal role in this domain (Durán-Muñoz & L'Homme, 2020).

Having said that, this investigation aims to analyse a type of grammatical collocation (i.e., verb + prepositional phrase) produced by motion verbs in a specialised corpus representing the language of adventure tourism. In particular, we explore two aspects: (1) the type of motion (real or fictive) represented by the collocations selected, and (2) the semantic roles belonging to the argument structures of the verbs expressed by the prepositions of the collocations. Ultimately, this study seeks to contribute to the characterisation of the language of adventure tourism by means of its phraseology. In the light of this, the organisation of this chapter is as follows. Section 2 tries to define the concept of collocation, and Section 3 examines the role of collocations in the specialised language of (adventure) tourism. After that, Section 4 describes the methodology applied to achieve our aims, and the main findings are presented in Section 5. Finally, Section 6 summarises the conclusions of this work and identifies new areas for further research.

2. Understanding the Concept of *Collocation*

As aforementioned, it was in the 1940s when scholars started to classify PUs. Their works inspired others and, since then, many have been the categorisations available within the field of phraseology. Dating back to the 1960s, N. N. Amosova (1963) accepted the ideas developed by her predecessor, that is, V. V. Vinogradov (1947). Although these two linguists did not use the term *collocation*, they noticed a type of combination where an element preserved its direct meaning and the other element was used figuratively (e.g., *small talk, small hours*). Vinogradov (1947) called these units *phraseological combinations*, and Amosova (1963) *phraseoloids*.

A couple of decades later, A. P. Cowie (1981) and P. A. Howarth (1996) suggested a categorisation of word combinations which was slightly different from that of their predecessors, given that these two authors were more interested in combination restrictions and strived to separate restricted from open expressions. In this case, they did use the term *collocation*, more specifically, *restricted collocation*, to refer to a combination in which one of the elements is interpreted literally while the other conveys a figurative meaning (e.g., *to capture somebody's imagination, to capture images*). Aside from that, they also employed the term *open collocation* to name a completely transparent combination in which both elements are read literally (e.g., *to drink tea, to drink water*).

Moving on to the twenty-first century, the development of the linguistic branches Natural Language Processing and Corpus Linguistics has influenced the categorisation of PUs. For instance, Sag et al. (2002, pp. 4–8) suggest a taxonomy of multi-word expressions (henceforth, MWEs) based on not only grammar rules but also statistics; therefore, *collocations* are said to be any statistically significant co-occurrence, regardless of their degree of flexibility or idiomaticity. Granger and Paquot (2008) aim to extract MWEs that can be either pervasive or infrequent in language; thus, *(lexical) collocations* are frequent combinations which are used referentially (e.g., *heavy rain, closely linked*). For their part, Baldwin and Kim (2010) use the term *light-verb construction* to refer to a PU which fulfils the requirement of not only statistical recurrence but also lexical, syntactic, semantic, and/or pragmatic idiomaticity (e.g., *to make an attempt, to take a bath*).

All in all, two perspectives regarding the concept of collocation must be distinguished. On the one hand, the criteria of structural flexibility and semantic opacity have been applied to define collocations as word combinations whose structure is partially flexible (i.e., their elements allow some substitution) and which are semantically transparent (i.e., one of the elements is interpreted literally, and the other one figuratively). On the other hand, the idea of a collocation as a frequent combination matching the criterion of statistical tendency arises from computational and corpus-assisted studies. Despite these two approaches, a set of features allows the distinction between collocations and other word combinations; the former are further explained in Section 2.1.

2.1. Defining the concept of (general and specialised) collocation

Principally, definitions of the term collocation have been formulated on the basis of several aspects. First, a collocation is normally defined as a combination of two words (Gablasova et al., 2017; Uhrig et al., 2018), namely base (Hausmann, 1979) or node (Sinclair, 1991) and collocate. Additionally, other elements might be found, although the base or node always enjoys a privileged status, given that it is freely chosen (or is the element under study) and determines the choice of the other (i.e., the collocate) (Sánchez-Berriel et al., 2018; Valencia Giraldo & Corpas Pastor, 2019).

Second, regarding the type of elements involved (Benson et al., 1986), two are the basics: lexical words (e.g., noun, adjective, verb, adverb) and function words (e.g., preposition, conjunction). Thus, a grammatical collocation is a combination of a lexical word with a preposition or grammatical structure, such as a *that*-clause or a *to*-infinitive (e.g., *an agreement that*, *necessary to*), while a lexical collocation is a combination of two lexical words, such as *warmest regards*, *to commit murder*.

Third and as aforementioned, collocations are partially opaque when it comes to their meaning, for one of the elements is interpreted figuratively and the other allows a literal reading. Added to that, their constituents are said to combine arbitrarily (Zaabalawi & Gould, 2017), given that collocations are often non-predictable (Benson et al., 2009) and do not conform to (at least semantic) rules. For this reason, they must be interpreted as *holistic units* (Revier, 2009, p. 128).

Next, a key aspect when defining the concept of collocation is that its elements do not necessarily co-occur uninterruptedly in the text. According to Sinclair (1991), we can find relevant collocations within a word span of -4/+4, which means four words on either side of the base. A similar view has been supported by other authors who have applied spans of -4/+4 (Lindquist, 2009; Mehler, 2009) and -5/+5 (Culpeper & Demmen, 2015; Evert, 2009; Patiño García, 2014) in their studies.

Last but not least, collocations are established on the basis of their frequency of use (Brezina et al., 2015; Cao & Deignan, 2019; Riches et al., 2021). Thanks to corpus linguistics, which relies on quantitative methods, it is possible to gather accurate objective information about the frequency of co-occurrence (one word co-occurring with another) and recurrence (the frequency of two words co-occurring together in a language) of words. Nevertheless, it must be highlighted that a consensus on how often two elements must co-occur to be regarded as a collocation has not been reached yet.

In spite of the fact that these characteristics have mainly been attributed to collocations found in the general language, they also match specialised collocations (L'Homme, 2017; Lorente Casafont, 2002; Lorente Casafont et al., 2017; Patiño García, 2014). To put it differently, collocations occurring in specialised languages are also combinations of two words (or better said two terms, or at least one term) which show a statistical tendency of preference. These elements do not need to be immediately adjacent to each other in the text, and the whole unit is unpredictable and semi-compositional. Moreover, a heavy emphasis is placed on the fact that a specialised collocation contains a special reference within a specialised subject field. In this study, collocations found in adventure tourism are targeted, which is the issue discussed in Section 3.

3. Collocations in the Specialised Language of (Adventure) Tourism

The "multifaceted character of the language of tourism" (Mănescu, 2020, p. 223) has given rise to a growing body of research on different features (cf. Baynat Monreal, 2017; Cappelli, 2006; Fuster-Márquez & Pennock-Speck, 2015; Rață & Petroman, 2012). In this work, the focus lies on the phraseology used in this domain, which has also been the priority guiding other studies. For instance, Corbacho Sánchez (2005) performs a contrastive study of collocations containing the bases *Tour(ismus)* ['tourism'], *Reise* ['trip'], and *Fahrt* ['trip'] in German and their translations into Spanish. He concludes that metaphors play a significant role in specialised phraseology, given that the collocations found for *Tour(ismus)* do not hold a relationship with tourism. The same author (Corbacho Sánchez, 2017) explores verb + noun collocations in German tourism and acknowledges that it is characterised by a broad range of specific collocations which give it the status of specialised language.

Another relevant study is carried out by Manca (2008), who aims to know the influence of culture and context of situation in phraseology. She uses a British farmhouse holiday corpus and centres on specialised collocations containing qualifying adjectives, which are considered the base. She states that adjectives can be classified into three semantic fields, namely, "description of rooms", "description of surroundings", and "description of food". A key point is that adjectives refer to "size" (e.g., *spacious, large*), "equipment" (e.g., *equipped, furnished*), and "beauty" (e.g., *attractive, beautiful*).

For their part, Piccioni and Pontrandolfo (2019) aim to identify and classify metaphorical collocations related to space in a bilingual (Spanish-Italian) corpus containing all text types used in tourism (e.g., guides, blogs, contracts). Moreover, their ultimate goal is to know how metaphorical phraseology can help tourists create an image of their tourist destination. Their results reveal that metaphors based on body parts are behind some collocations, for example, *pulmón verde* ['green lung'] or *arteria urbana* ['urban artery']. Besides that, metaphors are present in fictive motion verbs and represent a static entity as if it were moving, such as *la llanura manchega se alza* ['the plain of La Mancha rises up'] or *Un excelente y amplio golfo se extiende* ['An excellent and wide gulf stretches'], helping the tourist perceive the movement.

Collocations of motion verbs are also the topic of the work conducted by Durán-Muñoz and Jiménez-Navarro (2021). They present a contrastive study (English-Spanish) of collocations of verbs included in the semantic frame TRAVERSING (*cross, pass, traverse*) in the segment of adventure tourism. In total, they identify 23 collocations in Spanish and 17 collocations in English of three different types in this order: verb + noun, noun + verb, and verb + preposition. They claim that all the combinations examined are definitely specialised collocations, given that their combinatorial attraction is higher in the specialised corpus than in the reference corpus used in the analysis.

Finally, one more piece of work by these authors (Jiménez-Navarro & Durán-Muñoz, 2024) addresses the collocations produced by a list of motion verbs in adventure tourism. More specifically, they investigate 35 lexical collocations of (from more to less productive) *climb, cross, descend, ascend, head, wind, reach*, and *rush*, all of them describing fictive motion. Their main findings show that collocational production and strength are not correlated, since highly productive verbs do not produce really strong collocations. On the other hand, their syntactic analysis of these PUs sheds additional light on the semantic roles played by the collocates of the verbs.[1]

In short, these studies, particularly the two mentioned in the last place, are the background of the current investigation, whose overall aim is to contribute to the characterisation of the language of adventure tourism, particularly through the analysis of collocations of motion verbs plus prepositional phrases. Section 4 describes the methodology used to accomplish this goal.

---

1   Cappelli (this volume) also explores collocations in the language of adventure tourism, specifically, the collocational patterns of the nouns *person* and *people*.

4. Methodology

The methodology applied in this research was corpus-driven and relied on the specialised corpus of adventure tourism ADVENCOR (Durán-Muñoz & Jiménez-Navarro, 2021). This corpus was semi-automatically compiled using the Sketch Engine[2] tool and contains around one million words found in promotional texts published in English-speaking countries. Three steps were followed: (1) extraction and selection of motion verbs, (2) extraction of the collocations of the verbs selected, and (3) identification of the type of motion expressed by the collocations.

4.1. Extraction and selection of motion verbs

The *Keywords* function of Sketch Engine was used and it extracted a potential list of motion verbs which would be the bases of the grammatical collocations analysed in the current study. This function compares the frequency and relative frequency of the same words in the specialised corpus and a reference corpus using the keyness score *simple math* (Kilgarriff, 2009). No threshold was set at this point because simple math produces real keywords, although a minimum frequency of the verbs was set in two tokens, as it would be the minimum frequency for a word combination to be considered "a potential collocation" (Evert, 2009, p. 1215).

This step produced 978 candidate verbs, which were checked thoroughly to discard those elements that did not meet the selection criteria. As a consequence, 860 items were discarded for these reasons: (1) they were not motion verbs (e.g., *beep*, *enrol*), (2) they had been wrongly tagged (e.g., *NZD*, *pp*), (3) they were abbreviations (e.g., *yrs*, *lbs*), (4) they were not related to adventure tourism (e.g., *tinkle*, *allay*), or (5) they displayed distinct lemmas of the same motion verb (e.g., *paraglided, paraglides, paragliding* for *paraglide*). The final list of motion verbs amounted to 118.

4.2. Extraction of verb + prepositional phrase collocations

The *Word Sketch* function of Sketch Engine allowed the extraction of the collocations of the verbs previously selected. This function relies on statistical and linguistic criteria and allows the user to retrieve a one-page summary of the collocational preferences of terms, organising the results into columns according to the syntactic relationship between the term under study and its collocates. In this case, the focus was on prepositional phrases.

---

2 https://www.sketchengine.eu/ [Last accessed: 28/09/2023].

In order to make sure that free combinations were not extracted, two criteria were established. First of all, the minimum *logDice* score, which is the association measure used in the tool, was set in 7 points, given that this is half the theoretical maximum value of this score (i.e., 14 points); therefore, it could serve as a useful indicator of a significant association (Aldhubayi & Alyahya, 2014). On the other hand, a minimum frequency threshold was set in two tokens, for the best results are obtained by an association measure combined with a frequency threshold (Evert, 2009). Although this type of threshold is normally arbitrary, two tokens have proved to be valid in the specialised language of adventure tourism (cf. Jiménez-Navarro, 2020).

4.3. Identification of the type of motion expressed by the collocations

The last step of this methodology involved the identification of the type of motion represented by the collocations selected, that is, real motion, where there is a displacement of people in an adventure activity (e.g., *The jump is not compulsory, you can abseil down*[3]), or fictive motion, where an inanimate entity is described as if it were moving, but it is static in the real world (e.g., *A number of trails ascend to the top, including the most popular, the 13-mile Barr Trail*). This distinction was possible using the *Concordance* function of Sketch Engine, which shows the collocations in context. At this point, two more criteria were applied: first, two of the contexts selected for each collocation should represent the same type of motion, and second, these contexts should contain a specific reference to the domain of adventure tourism.

5. Results

As it was revealed in Section 4.1., the number of motion verbs selected was 118, demonstrating the rich inventory of this type of verb used in the domain of adventure tourism. Nevertheless, only half of them (48 %) produced collocations satisfying all the extraction and selection criteria. The Appendix shows this list of verbs in alphabetical order along with the number of collocations they produced and the type of motion described. The most productive verbs were *fly*-v (22 collocations), *walk*-v (18), *climb*-v (16), *jump*-v (16), *soar*-v (14), *travel*-v (13), and *hike*-v (11), but only *climb*-v produced collocations representing both

---

3 All the examples related to adventure tourism were extracted from the ADVENCOR corpus.

real motion (13 collocations) and fictive motion (three) (the other verbs only produced real motion collocations). In total, 273 verb + prepositional phrase collocations were extracted from the corpus, 249 of them describing real motion, and 24 of them describing fictive motion.

In most cases, those collocations displaying fictive motion belong to motion verbs which also produced collocations representing real motion, and are seven verbs in total: *ascend*-v, *climb*-v, *descend*-v, *dive*-v, *fall*-v, *pass*-v, and *run*-v. However, three other verbs must be highlighted, since they only produced collocations displaying fictive motion events: *drop*-v (one collocation), *meander*-v (one), and *wind*-v (two). On the whole, 82 % of the motion verbs produced collocations which revealed only real motion events, 13 % of them produced collocations which showed both types of motion events, and 5 % of them produced collocations which illustrated only fictive motion events.

Section 5.1. examines those collocations representing only real motion, whereas Section 5.2. explores those collocations describing fictive motion. These analyses are performed in terms of the argument structures of the verbs, given that they aim to explain which semantic roles belonging to those structures (either arguments, i.e., elements that are central to the meaning of the verbs, such as Destination, Path, or Source, or circumstantials, i.e., elements that are peripheral to the meaning of the verbs and optional when characterising it, such as Distance, Method, or Purpose) are evoked by the collocations, or better said by the prepositions.

5.1. Verb + prepositional phrase collocations representing real motion

In this work, 91 % of the collocations extracted describe real motion events. In general, the entity undergoing displacement denotes a human being participating in an adventure activity (247 collocations), except for two collocations which show an animal (*sled* + *through* + place, e.g., *As your sled-dogs sled through the frozen lake [...]*) and a vehicle (*fly* + *at* + *(an) altitude*, e.g., *[...] as you have to jump from an aircraft flying at a higher altitude*) moving.

At this point, two types of collocations are distinguished: on the one hand, collocations in which the object following the preposition is specified, for example, *arrive* + *at* + *(a) destination*, *float* + *down* + *(a) river*, *paddle* + *in* + *(a) kayak*; on the other hand, collocations in which the object is unspecified, for instance, *cross* + *over* + place, *jump* + *on* + place, *walk* + *for* + time. This distinction arises from the extraction options offered in Sketch Engine, but it has no impact on the results of this study because the key element of the collocation is the preposition.

As aforementioned, the most productive verbs of collocations representing real motion are *fly*-v (22 collocations), *walk*-v (18), *jump*-v (16), *soar*-v (14), *climb*-v (13), *travel*-v (13), and *hike*-v (11). They are verbs which imply distinct sorts of meaning, as it can be observed in the following list, which shows a classification of the verbs according to the meaning encoded by themselves and their percentage distribution in the data:[4]

1. Verbs describing motion which follows a non-specified trajectory (20 %): *explore*-v, *float*-v, **hike-v**, *hop*-v, *journey*-v, *run*-v, *sail*-v, **travel-v**, *trek*-v, *venture*-v, **walk-v**.
2. Verbs encoding vertical motion, either upward or downward DIRECTION of motion (18.5 %): *abseil*-v, *ascend*-v, **climb-v**, *descend*-v, *fall*-v, *plunge*-v, *rappel*-v, *scramble*-v, *slide*-v, **soar-v**.
3. Verbs focused on the VEHICLE used for motion (17 %): *bike*-v, *cruise*-v, *cycle*-v, *drive*-v, *navigate*-v, *paddle*-v, *ride*-v, *sled*-v, *zip*-v.
4. Verbs oriented towards the SOURCE from which the motion originates (13 %): *depart*-v, *exit*-v, **jump-v**, *launch*-v, *leap*-v, *skydive*-v, *transfer*-v.
5. Verbs oriented towards the DESTINATION of motion (11 %): *arrive*-v, *disembark*-v, *enter*-v, *head*-v, *land*-v, *reach*-v.
6. Verbs oriented towards the PLACE where the motion occurs (11 %): *dive*-v, **fly-v**, *glide*-v, *snorkel*-v, *swim*-v, *wade*-v.
7. Verbs focused on the SPEED of motion (5.5 %): *accelerate*-v, *race*-v, *speed*-v.
8. Verbs focusing attention on a PATH (4 %): *cross*-v, *pass*-v.

In addition to the meaning evoked by the verbs, in several collocations the prepositional collocate is also a signal of that meaning. For instance, the prepositions *from* and *off* were found in seven (out of 16) collocations produced by the verb *jump*-v (e.g., *jump + from + (a) plane, jump + off + (a) bridge*), which express motion starting from a SOURCE towards a DESTINATION (this last semantic role is described by four collocations too, e.g., *jump + into + (a) pool*). In the same vein, some verbs representing motion on a vertical axis co-occur with prepositions emphasising the DIRECTION of motion, such as *abseil + down + (a) waterfall, climb + up +* place. Nevertheless, the perfect example of this is the case of verbs evoking DESTINATION, given that 19 out of 23 collocations produced

---

4 The most productive verbs of collocations representing real motion are highlighted in bold.

by these verbs contain a preposition which semantically encodes this role when combined with those bases, such as *arrive + at + (a) bridge, enter + into +* place. Examples (1)–(3)[5] show contexts where some of these collocations occur.

(1) We **jumped** *from the plane*, did a couple flips, and began our free-fall for about 40 seconds. [SOURCE]
(2) Discover the incredible natural beauty of the Blue Mountains as you **abseil** *down waterfalls* […]. [DIRECTION]
(3) Also bring dry clothes to wear once we **arrive** *at the bridge*. [DESTINATION]

As for the recurrence of the prepositions collocating with the verbs, Figure 1 shows that *through* is the most frequent preposition (44 collocations), followed at a considerable distance by *in* (27), *to* (26), *at* (19), *down* (18), *into* (18), and *from* (15). At the bottom of the list are: first, *toward/s*, found in five collocations (e.g., *drive/head/trek + towards +* place); second, *along, around,* and *with*, identified in four collocations each (e.g., *travel + along +* place, *paddle + around +* place, *hike + with + (a) group*); third, *above, by,* and *like*, each of them found in three collocations (e.g., *float + above +* place, *pass + by +* vehicle, *fly + like + (a) bird*); fourth, *under*, which is identified in two collocations (i.e., *paddle + under +* place and *swim + under + (a) waterfall*); finally, *as, before, past,* and *throughout*, each of them identified in one collocation (i.e., *climb + as + (a) group, arrive + before + time, cruise + past +* place, *travel + throughout +* place).

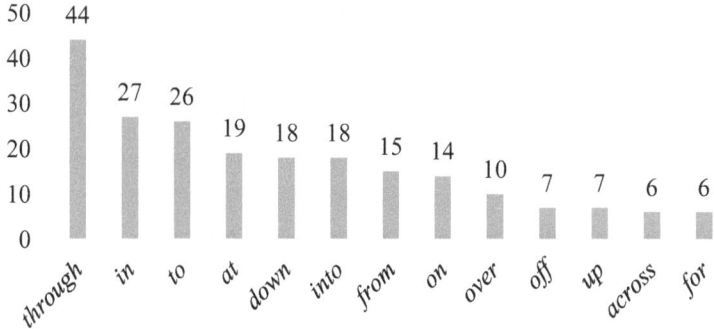

Figure 1. Frequency of the prepositions identified in collocations representing real motion

---

5 The verbs of the collocations are highlighted in bold, and the prepositional phrases combining with them are in italics.

In all the examples, *through* introduces the PATH where the motion is carried out, as represented in example (4). About the preposition *in*, it introduces the PLACE where motion occurs in most of the collocations (70 %; example (5)), although it can also introduce the VEHICLE_WITH_ENGINE or the VEHICLE_WITHOUT_ENGINE used in the motion (15 %; examples (6) and (7), respectively). Regarding *to*, it introduces DESTINATION in 96 % of the collocations (example (8)) and SPEED in only one collocation (example (9)). Moving on to *at*, it can describe different semantic roles, such as DESTINATION (42 %; example (3)), PLACE (26 %; example (10)), SPEED (26 %; example (11)), and DISTANCE (6 %; example (12), although here the subject performing the motion is a vehicle, i.e., an aircraft). On the other hand, *down* always represents downward DIRECTION of motion (example (2)), and *into* introduces the semantic role DESTINATION (56 %; example (13)). Finally, *from* always introduces the SOURCE of the motion (example (1)).

(4) **Soar** *through the clouds*, feeling free as a bird […]. [PATH]
(5) The highlights of this active holiday include canoeing/kayaking, rafting, hiking and **cycling** *in the area around Plitvice Lakes National Park*. [PLACE]
(6) The experience of **flying** *in a helicopter* is really astounding! [VEHICLE_WITH_ENGINE]
(7) **Paddle** out *in our kayaks* and explore the wonderful ecosystem that surrounds this area. [VEHICLE_WITHOUT_ENGINE]
(8) We will **bike** *to the shores of Visovac Lake* […]. [DESTINATION]
(9) […] you'll **accelerate** *to speeds of up to 220 KPH (124 MPH)*. [SPEED]
(10) Kayakers may find it easier to **launch** *at the less congested ramps* […]. [PLACE]
(11) You can enjoy nature, work your body and **explore** *at your own pace*. [SPEED]
(12) […] you have to jump from an aircraft **flying** *at a higher altitude*. [DISTANCE]
(13) Then you can traverse up our sky bridge to **enter** *into the canopy of the trees*. [DESTINATION]

5.2. Verb + prepositional phrase collocations representing fictive motion

In this study, 9 % of the collocations extracted represent fictive motion events. In some cases, this motion is perceived by the human eye as such because the observer is moving through the landscape, as in *The path is rising quickly as we climb* (Langacker, 2005, p. 175). However, this type of example, in which the human being is experiencing the movement, was not extracted from ADVENCOR, but we retrieved other contexts where no people are mentioned, such as *More than 20 miles of hiking trails that climb onto spectacular views […]*.

In total, 24 collocations representing fictive motion were identified. The number of verbs producing this type of collocation was 10, and the most productive ones were *run*-v (six collocations), *climb*-v (three), *descend*-v (three), and *pass*-v (three), all of which also produced real motion collocations. In fact, in several cases the same collocation was found to describe both types of motion, such as *ascend* + *to* + place,[6] *descend* + *into* + place,[7] and *pass* + *through* + place.[8] For their part, the verbs producing fictive motion collocations are classified according to their implied meaning as follows:[9]

1. Verbs encoding vertical motion, either upward or downward DIRECTION of motion (46 %): *ascend*-v, **climb-v**, **descend-v**, *drop*-v, *fall*-v.
2. Verbs describing motion which follows a non-specified trajectory (PATH) (18 %): **run-v**.
3. Verbs focused on the shape of the path (18 %): *meander*-v, *wind*-v.
4. Verbs oriented towards the PLACE where the motion occurs (9 %): *dive*-v.
5. Verbs focusing attention on a PATH (9 %): **pass-v**.

Unlike real motion collocations, in which the preposition contributes to the meaning encoded by the base, in fictive motion collocations it is not usual. For example, the prepositions accompanying the vertical motion verbs do not introduce the DIRECTION of motion but rather its DESTINATION, such as *ascend* + *to* + place (e.g., *Today the trail* ***ascends*** *first to a ridge [...]*) and *descend* + *into* + place (e.g., *Stay with the trail as it* ***descends*** *into a winding wash*). The most common semantic role expressed by the prepositions is PATH (12 collocations), followed by DESTINATION (10), PLACE (one), and SOURCE (one). Therefore, in contrast to real motion collocations, whose prepositions also describe circumstantials, that is, semantic roles which are less central to the meaning of

---

6   Real motion example: *On the way back to Narsarsuaq we* ***ascend*** *to the highest peak in Mellem Land to enjoy one of the most exclusive views of southern Greenland*; fictive motion example: *Today the trail* ***ascends*** *first to a ridge before dropping steeply through Simbu (1700m)*.
7   Real motion example: *It feels like you're* ***descending*** *into the bowels of some gigantic prehistoric beast, a sensory feast*; fictive motion example: *Stay with the trail as it* ***descends*** *into a winding wash*.
8   Real motion example: *You will fly over cenotes and* ***pass*** *through the caves of Xel-Há*; fictive motion example: *The road* ***passed*** *through a national park gate where the vehicle was inspected*.
9   The most productive verbs of collocations representing fictive motion are highlighted in bold.

the verbs (e.g., DURATION, MANNER, SPEED), the prepositions found in fictive motion collocations only represent arguments. Examples (14)–(21) show some contexts.[10]

(14) The terrain includes <u>footpaths</u> that **meander** *through the lush secondary forest that leads to lookout points.* [PATH]
(15) Despite the misty weather, we pushed through <u>our first via ferrata of the trip – the Maximiliansteig</u>, which **runs** *along the ridge of Cima di Terrarossa.* [PATH]
(16) <u>The main canyon</u> actually **runs** *alongside the road* [...]. [PATH]
(17) <u>The trail</u> **drops** *to the grassy pass overlooking a huge landslip* with a view of Mt. Jannu (7,711m) [...]. [DESTINATION]
(18) <u>The path</u> [...] then slowly **descends** *into the forest.* [DESTINATION]
(19) <u>More than 20 miles of hiking trails</u> that **climb** *onto spectacular views* and weave alongside clear streams and waterfalls. [DESTINATION]
(20) [...] <u>the second pitch</u> **climbs** *over a 60-foot pillar and a few shorter ones* to the top of the route. [PLACE]
(21) <u>A second branch of the trail</u> **runs** *from north of Horseshoe Valley* to Highway 32 [...]. [SOURCE]

Regarding the frequency of the prepositions, *through* is also the most common one (10 collocations, 42 %; example (14)); in fact, it doubles the second on the list, which is *to* (21 %). *Through* is the only preposition together with *along* and *alongside* (one collocation each; examples (15) and (16)) representing the PATH along which the motion seems to occur. After that, *to* co-occurs in five collocations and describes the DESTINATION of the motion (example (17)). This semantic role is also expressed by *into* (four collocations; example (18)) and *onto* (one collocation; example (19)). Finally, the PLACE where the motion occurs is found in the only collocation with *over* (example (20)), and the SOURCE from which it originates is also detected in just one collocation, in this case containing *from* (example (21)). This prepositional distribution is shown in Figure 2. Finally, among the most recurrent stationary entities which seem to be moving are *trail* (found in 50 % of the contexts extracted), *path* (9 %), *road* (7.5 %), *pitch*, and *trek* (6 % each).

---

10 The subjects of the verbs (i.e., the static entities that seem to be moving) are underlined.

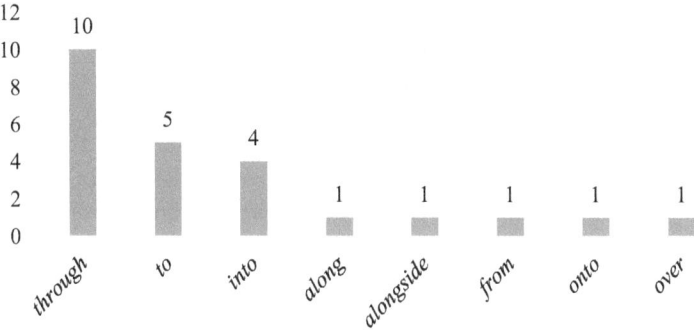

Figure 2. Frequency of the prepositions identified in collocations representing fictive motion

6. Conclusions

This study has explored a type of grammatical collocation, that is, verb + prepositional phrase, of a set of motion verbs extracted from a specialised corpus containing promotional texts about adventure tourism. In total, 118 motion verbs were first selected, but only 57 of them (cf. Appendix) produced collocations meeting all eligibility criteria: (1) the minimum *logDice* score of the collocation was 7 points, (2) the collocation occurred at least two times in the corpus and represented the same type of motion (real or fictive) in these two contexts, and (3) the contexts selected contained a specific reference to the domain of adventure tourism. After that, 273 collocations were selected, 91 % (249) of them showing real motion events and 9 % (24) of them depicting fictive motion events.

The results of this work have been presented in terms of the type of motion described by the collocations. As for the most productive verbs, *fly*-v (22 collocations), *walk*-v (18), *jump*-v (16), *soar*-v (14), and *climb*-v (13) were the top five in real motion events, whereas *run*-v (six collocations), *climb*-v (three), *descend*-v (three), *pass*-v (three), *ascend*-v (two), *dive*-v (two), and *wind*-v (two) were found in fictive motion events. From these lists, the following verbs produced collocations representing both types of motion: *ascend*-v, *climb*-v, *descend*-v, *dive*-v, *pass*-v, and *run*-v; even in some cases, the same collocation was found in contexts of real and fictive motion, such as *pass* + *through* + place.

One of the aspects of this analysis has referred to the semantic roles belonging to the argument structures of the verbs evoked by the prepositions co-occurring with the motion verbs. Thus, it has been discovered that in some real motion collocations the prepositions contribute to the meaning implied by the very verb, such as *jump* + *from* + place, given that *jump*-v is oriented towards the SOURCE of the motion and *from* introduces this element. Nevertheless, this was not found to be a feature of fictive motion collocations. On the other hand, while prepositions always represent arguments of the verbs in the latter (fictive motion), they can describe both arguments and circumstantials in the former (real motion). Regarding the prepositional frequency, *through* was the most recurrent preposition in both types of motion (20 % and 42 %, respectively); in fact, it almost doubled or even doubled the second preposition on the lists. As for the most common semantic role, it was PATH in the collocations representing real motion as well as fictive motion.[11]

Our goals have therefore been successfully achieved. Thus, one type of grammatical collocation (i.e., verb + prepositional phrase) produced by a selected list of motion verbs has been explored regarding two factors, the type of motion represented and the semantic roles expressed by the prepositions. As a result, this analysis has contributed to the characterisation of the language of adventure tourism, particularly, its phraseology, and has demonstrated the value of a corpus-driven methodology for such a study. The next step should lead to the implementation of the collocations extracted into the online specialised resource *DicoAdventure*[12] (cf. Jiménez-Navarro, 2020), and future studies should focus on contrastive studies (English-Spanish) of the same type of collocation, as well as on verb + prepositional phrase collocations produced by other sorts of verbs.

References

Aldhubayi, L., & Alyahya, M. (2014). Automated Arabic antonym extraction using a corpus analysis tool. *Journal of Theoretical and Applied Information Technology*, *70*(3), 422–433.

Amosova, N. N. (1963). *Osnovy anglijskoj frazeologii*. Izdatelstvo Leningradskogo Universyteta.

Baldwin, T., & Kim, S. N. (2010). Multiword expressions. In N. Indurkhya & F. J. Damerau (Eds.), *Handbook of Natural Language Processing* (second edition) (pp. 267–292). Taylor & Francis Group.

---

11 Interestingly, PATH is also a significant semantic role in other type of word combinations, more specifically, compounds with a head noun ending in -*ing*, where this role connects both constituents of the compound (cf. Portero Muñoz, this volume).
12 http://olst.ling.umontreal.ca/dicoadventure/ [Last accessed: 28/09/2023].

Baynat Monreal, M. E. (2017). El léxico de la gestión turística en lengua francesa en el Diccionario Multilingüe de Turismo: Análisis contrastivo con la lengua inglesa. *Çédille, Revista de Estudios Franceses, 13*, 53–82.

Benson, M., Benson, E., & Ilson, R. (1986). *The BBI Combinatory Dictionary of English: A guide to word combinations*. John Benjamins Publishing Company.

Benson, M., Benson, E., & Ilson, R. (2009). *The BBI Combinatory Dictionary of English: Your guide to collocations and grammar* (third edition). John Benjamins Publishing Company.

Brezina, V., McEnery, T., & Wattam, S. (2015). Collocations in context: A new perspective on collocation networks. *International Journal of Corpus Linguistics, 20*(2), 139–173. https://doi.org/10.1075/ijcl.20.2.01bre

Cao, D., & Deignan, A. (2019). Using an online collocation dictionary to support learners' L2 writing. In C. Wright, L. Harvey, & J. Simpson (Eds.), *Voices and practices in applied linguistics. Diversifying a discipline* (pp. 233–249). White Rose University Press.

Cappelli, G. (2006). *Sun, sea, sex and the unspoilt countryside. How the English language makes tourists out of readers*. Pari Publishing.

Cappelli, G. (This volume). The language of accessible adventure tourism. In I. Durán-Muñoz & E. L. Jiménez-Navarro (Eds.), *Exploring the language of adventure tourism: A corpus-assisted approach*. Peter Lang.

Corbacho Sánchez, A. (2005). Los conceptos de *Tour(ismus)* y *Reise* en los fraseologismos alemanes y su traducción al español. *Pragmalingüística, 13*, 65–75. https://doi.org/10.25267/Pragmalinguistica.2005.i13.04

Corbacho Sánchez, A. (2017). Colocaciones verbo-nominales de la correspondencia hotelera en alemán: Incidencia en la enseñanza y traducción al español. *Encuentro, 26*, 94–112.

Cowie, A. P. (1981). The treatment of collocations and idioms in learners' dictionaries. *Applied Linguistics, 2*(3), 223–235. https://doi.org/10.1093/applin/II.3.223

Culpeper, J., & Demmen, J. (2015). Keywords. In D. Biber & R. Reppen (Eds.), *The Cambridge handbook of English corpus linguistics* (pp. 90–105). Cambridge University Press.

Durán-Muñoz, I., & Jiménez-Navarro, E. L. (2021). Colocaciones verbales en el turismo de aventura: Estudio contrastivo inglés-español. In G. Corpas Pastor, M.ª R. Bautista Zambrana, & C. M. Hidalgo-Ternero (Eds.), *Sistemas fraseológicos en contraste: Enfoques computacionales y de corpus* (pp. 121–142). Comares.

Durán-Muñoz, I., & L'Homme, M.-C. (2020). Diving into English motion verbs from a lexico-semantic approach. A corpus-based analysis of adventure tourism. *Terminology*, *26*(1), 33–59. https://doi.org/10.1075/term.00041.dur

Evert, S. (2009). Corpora and collocations. In A. Lüdeling & M. Kytö (Eds.), *Corpus linguistics. An international handbook, vol. II* (pp. 1212–1248). Walter de Gruyter. https://doi.org/10.1515/9783110213881.2.1212

Fuster-Márquez, M., & Pennock-Speck, B. (2015). Target frames in British hotel websites. *IJES*, *15*(1), 51–69. https://doi.org/10.6018/ijes/2015/1/213231

Gablasova, D., Brezina, V., & McEnery, T. (2017). Collocations in corpus-based language learning research: Identifying, comparing, and interpreting the evidence. *Language Learning*, *67*(S1), 155–179. https://doi.org/10.1111/lang.12225

Granger, S., & Paquot, M. (2008). Disentangling the phraseological web. In S. Granger & F. Meunier (Eds.), *Phraseology. An interdisciplinary perspective* (pp. 27–49). John Benjamins Publishing Company.

Hausmann, F. J. (1979). Un dictionnaire des collocations est-il possible? *Travaux de Linguistique et de Littérature*, *17*(1), 187–195.

Howarth, P. A. (1996). *Phraseology in English academic writing. Some implications for language learning and dictionary making*. Niemeyer.

Jiménez-Navarro, E. L. (2020). *Treatment and representation of verb collocations in the specialised language of adventure tourism* [Doctoral dissertation, Universidad de Córdoba]. Helvia. https://helvia.uco.es/xmlui/handle/10396/20976

Jiménez-Navarro, E. L., & Durán-Muñoz, I. (2024). Collocations of fictive motion verbs in adventure tourism: A corpus-based study of the English language. *Revista Española de Lingüística Aplicada* (online). https://doi.org/10.1075/resla.21042.jim

Kilgarriff, A. (2009). Simple maths for keywords. In M. Mahlberg, V. González-Díaz, & C. Smith (Eds.), *Proceedings of the Corpus Linguistics Conference 2009 (CL2009)* (pp. 1–6). University of Liverpool.

Langacker, R. W. (2005). Dynamicity, fictivity, and scanning: The imaginative basis of logic and linguistic meaning. In D. Pecher & R. A. Zwaan (Eds.), *Grounding cognition: The role of perception and action in memory, language, and thinking* (pp. 164–197). Cambridge University Press.

L'Homme, M.-C. (2017). Combinatoire spécialisée: Trois perspectives et des enseignements pour la terminologie. *TTR*, *30*(1-2), 215–241. https://doi.org/10.7202/1060025ar

Lindquist, H. (2009). *Corpus linguistics and the description of English*. Edinburgh University Press.

Lorente Casafont, M. (2002). Terminología y fraseología especializada: Del léxico a la sintaxis. In G. Guerrero Ramos & M. F. Pérez Lagos (Eds.), *Panorama actual de la terminología* (pp. 159–179). Comares.

Lorente Casafont, M., Martínez-Salom, A., Santamaría, I., & Vargas Sierra, C. (2017). Specialised collocations in specialised dictionaries. In S. Torner Castells & E. Bernal (Eds.), *Collocations and other lexical combinations in Spanish. Theoretical, lexicographical and applied perspectives* (pp. 200–222). Routledge.

Manca, E. (2008). From phraseology to culture: Qualifying adjectives in the language of tourism. *International Journal of Corpus Linguistics*, *13*(3), 368–385. https://doi.org/10.1075/ijcl.13.3.07man

Mănescu, O. M. (2020). The complexity of the language of tourism. *British and American Studies*, *26*, 221–229.

Mehler, A. (2009). Large text networks as an object of corpus linguistic studies. In A. Lüdeling & M. Kytö (Eds.), *Corpus linguistics. An international handbook*, vol. I (pp. 328–382). Walter de Gruyter.

Patiño García, P. (2014). Towards a definition of specialised collocation. In G. Quiroz Herrera & P. Patiño García (Eds.), *LSP in Colombia: Advances and challenges* (pp. 119–133). Peter Lang.

Piccioni, S., & Pontrandolfo, G. (2019). La construcción del espacio turístico a través de la fraseología metafórica. *Linguistik Online*, *94*(1/19), 137–153. https://doi.org/10.13092/lo.94.5436

Portero Muñoz, C. (This volume). The use of compounds in the adventure tourism lexicon. In I. Durán-Muñoz & E. L. Jiménez-Navarro (Eds.), *Exploring the language of adventure tourism: A corpus-assisted approach*. Peter Lang.

Rață, G., & Petroman, I. (2012). English borrowings in the Romanian of tourism (sites of travel agencies). In G. Rață, I. Petroman, & C. Petroman (Eds.), *The English of tourism* (pp. 49–54). Cambridge Scholars Publishing.

Revier, R. L. (2009). Evaluating a new test of whole English collocations. In A. Barfield & H. Gyllstad (Eds.), *Researching collocations in another language: Multiple interpretations* (pp. 125–138). Palgrave Macmillan.

Riches, N., Letts, C., Awad, H., Ramsey, R., & Dąbrowska, E. (2021). Collocational knowledge in children: A comparison of English-speaking monolingual children, and children acquiring English as an Additional Language. *Journal of Child Language*, 1–16. https://doi.org/10.1017/S0305000921000490

Sag, I. A., Baldwin, T., Bond, F., Copestake, A., & Flickinger, D. (2002). Multiword expressions: A pain in the neck for NLP. In A. F. Gelbukh (Ed.), *Lecture Notes in Computer Science. Volume 2276. Computational Linguistics and Intelligent Text Processing* (pp. 1–15). Springer.

Sánchez-Berriel, I., Santana Suárez, O., Gutiérrez Rodríguez, V., & Pérez Aguiar, J. R. (2018). Network analysis techniques applied to dictionaries for identifying semantics in lexical Spanish collocations. In P. Cantos-Gómez & M. Almela-Sánchez (Eds.), *Lexical collocation analysis: Advances and applications* (pp. 39–57). Springer. https://doi.org/10.1007/978-3-319-92582-0_3

Sinclair, J. (1991). *Corpus, concordance, collocation*. Oxford University Press.

Uhrig, P., Evert, S., & Proisl, T. (2018). Collocation candidate extraction from dependency-annotated corpora: Exploring the differences between parsers and dependency annotation schemes. In P. Cantos-Gómez & M. Almela-Sánchez (Eds.), *Lexical collocation analysis: Advances and applications* (pp. 111–140). Springer. https://doi.org/10.1007/978-3-319-92582-0_6

Valencia Giraldo, M. V., & Corpas Pastor, G. (2019). *The Portrait of Dorian Gray*: A corpus-based analysis of translated verb + noun (object) collocations in Peninsular and Colombian Spanish. In G. Corpas Pastor & R. Mitkov (Eds.), *Computational and corpus-based phraseology. Third International Conference, Europhras 2019, Malaga, Spain, September 25–27, 2019. Proceedings. Lecture Notes in Artificial Intelligence 11755* (pp. 417–430). Springer. https://doi.org/10.1007/978-3-030-30135-4_30

Vargas Sierra, C. (2012). La tecnología de corpus en el contexto profesional y académico de la traducción y la terminología: Panorama actual, recursos y perspectivas. In M. A. Candel Mora & E. Ortega Arjonilla (Eds.), *Tecnología, traducción y cultura* (pp. 67–99). Tirant Humanidades.

Vinogradov, V. V. (1947). *Izbrannye Trudy. Leksikologija i leksikografija*. Nauka.

Zaabalawi, R. S., & Gould, A. M. (2017). English collocations: A novel approach to teaching the language's last bastion. *Ampersand, 4*, 21–29. https://doi.org/10.1016/j.amper.2017.03.002

Acknowledgements

This chapter has been written thanks to support from the R&D project "DicoAdventure: diseño y desarrollo de un recurso electrónico especializado bilingüe (inglés, español) sobre el turismo de aventura a partir de marcos semánticos" (Ref. UCO-1380857-F), co-funded by the Operational Programme FEDER 2014–2020 and the Consejería de Economía, Conocimiento, Empresas y Universidad of the Andalusian regional government.

Appendix: Motion verbs which produced collocations and type of motion represented

Verb	Number of collocations produced	Real motion	Fictive motion
1. abseil-v	4	4	-
2. accelerate-v	1	1	-
3. arrive-v	8	8	-
4. ascend-v	3	1	2
5. bike-v	1	1	-
6. climb-v	16	13	3
7. cross-v	4	4	-
8. cruise-v	1	1	-
9. cycle-v	1	1	-
10. depart-v	3	3	-
11. descend-v	7	4	3
12. disembark-v	1	1	-
13. dive-v	4	2	2
14. drive-v	8	8	-
15. drop-v	1	-	1
16. enter-v	1	1	-
17. exit-v	1	1	-
18. explore-v	2	2	-
19. fall-v	8	7	1
20. float-v	9	9	-
21. fly-v	22	22	-
22. glide-v	6	6	-
23. head-v	7	7	-
24. hike-v	11	11	-
25. hop-v	1	1	-
26. journey-v	1	1	-
27. jump-v	16	16	-
28. land-v	5	5	-
29. launch-v	4	4	-
30. leap-v	2	2	-
31. meander-v	1	-	1
32. navigate-v	1	1	-

Verb	Number of collocations produced	Real motion	Fictive motion
33. paddle-v	9	9	-
34. pass-v	7	4	3
35. plunge-v	1	1	-
36. race-v	1	1	-
37. rappel-v	2	2	-
38. reach-v	1	1	-
39. ride-v	7	7	-
40. run-v	9	3	6
41. sail-v	1	1	-
42. scramble-v	3	3	-
43. skydive-v	2	2	-
44. sled-v	1	1	-
45. slide-v	2	2	-
46. snorkel-v	1	1	-
47. soar-v	14	14	-
48. speed-v	1	1	-
49. swim-v	7	7	-
50. transfer-v	1	1	-
51. travel-v	13	13	-
52. trek-v	4	4	-
53. venture-v	2	2	-
54. wade-v	1	1	-
55. walk-v	18	18	-
56. wind-v	2	-	2
57. zip-v	2	2	-
	273	249	24

Marie-Claude L'Homme

Observatoire de linguistique Sens-Texte (OLST),
Université de Montréal
mc.lhomme@umontreal.ca

7 Frame Semantics and domain-specific resources

Abstract: Designers of domain-specific resources have access to different methods and tools to identify terms and collect information from specialised texts as well as to various models to describe terms and represent relations between them. These methods and models (and the theoretical frameworks on which they are based) can raise questions about the linguistic content of resources, four of which are examined in this chapter: (1) kinds of terms to be taken into consideration, (2) the linguistic behaviour of terms, (3) capturing semantically related terms in ways that are meaningful to users of resources, and (4) connecting linguistic behaviour to knowledge. In this chapter, answers provided by Frame Semantics (Fillmore, 1976, 1982; Fillmore & Baker, 2010) and FrameNet (2023; Ruppenhofer et al., 2016) are examined as well as concrete implementations in two resources, namely *DicoAdventure* (2023), a resource that records terms in the field of adventure tourism, and *DiCoEnviro* (2022), a resource that contains environment terms.

Keywords: adventure tourism, environment, Frame Semantics, predicative term, terminology

1. Introduction[1]

Designers of domain-specific (or terminological) resources must take into consideration different kinds of requirements which all present their own challenges. They (often, terminologists or specialised lexicographers) can be requested to compile resources in very different domains in which they have no prior training. Resources can record terms in broad domains (e.g., environment or tourism) or, rather, focus on narrower topics (e.g., adventure tourism vs tourism, endangered species instead of environment). Resources can also be targeted at different kinds of users who have specific information needs

---

1 This chapter is the revised version of a presentation given at the Traditur Conference, at Universidad de Córdoba (Spain) in October 2022.

(Bergenholtz & Tarp, 1995), such as translators, knowledge modellers, or future experts.

These factors and other ones compel designers to make decisions that inevitably have an impact on the content of resources. Below are some of the questions designers need to tackle in the course of their work. Of course, more questions can arise in the process, but these are the ones that are addressed in this chapter:

1. What kinds of terms should be taken into consideration? Should different kinds of concepts (entities, events, properties, etc.) or, more specifically, terms that express these concepts be considered? Should the resource record nouns only or also terms that belong to other parts of speech?
2. Do terms behave differently? If nouns, verbs, adjectives, and adverbs or terms that designate entities, events, and properties are recorded, how should their specific linguistic properties be taken into consideration?
3. Which terms are related and how? How can related terms be identified? Which kinds of relations should be taken into consideration? How can relations be represented?
4. Can a resource represent both knowledge structure and the linguistic properties of terms? Can a connection be made between these two representation levels? Since terms express knowledge in given domains, how can we link a description that accounts for detailed linguistic properties to the knowledge structures of specialised domains?

Terminologists and other designers of domain-specific resources can use different methodologies and tools to locate and collect information from corpora. This first set of methods and tools give access to data in running text and usually require that a validation be made by a terminologist, an expert, or both. Designers also have access to various models for presenting terms and accounting for relations they share with one another. These models can provide different perspectives on terminological data depending on the needs of users.

Designers can also base their approach to term description on specific theoretical frameworks. Over the past few decades, frameworks to which terminologists refer have become more diversified. This chapter focuses on Frame Semantics (Fillmore, 1976, 1982; Fillmore & Baker, 2010) as well as the methodology that was developed to compile FrameNet[2] (Ruppenhofer et al., 2016). Although Frame Semantics was not originally devised for specialised

---

2   https://framenet.icsi.berkeley.edu/ [Last accessed: 28/09/2023].

knowledge or terms that convey it, it is increasingly mentioned in terminology literature. This chapter examines how the framework can be helpful when describing terms and, more specifically, how it can:

1. Help account for terms that are often overlooked in domain-specific resources (namely, verbs and adjectives) and their specific linguistic behaviour.
2. Help identify related terms, particularly terms that are linked to conceptual situations.
3. Allow terminologists to connect the linguistic properties of terms to a representation of the knowledge structure of a domain.

Two examples are used throughout the chapter to illustrate how principles borrowed from Frame Semantics and FrameNet can be implemented in domain-specific resources.[3] The first resource, called *DicoAdventure*[4] (Durán-Muñoz & L'Homme, 2020) records terms in the domain of adventure tourism and currently contains English and Spanish verbs, in addition to a few nouns denoting adventure activities. The second resource is an online dictionary called *DiCoEnviro*[5] combined with a frame layer (*Framed DiCoEnviro*;[6] L'Homme et al., 2020) which focuses on the environment and records nouns, verbs, adjectives, and adverbs in Chinese (Zheng, 2021), English, French, Portuguese (Lamberti Arraes, 2022), and Italian.[7] Examples given in this chapter are taken from the English versions of the resources.

The chapter is organised as follows: Section 2 examines a few questions raised by the fields of adventure tourism and the environment from the viewpoint of terminologists who must compile resources in these domains. Section 3

---

3 Two specific resources are mentioned in this chapter but other projects have applied the methodology developed for FrameNet to different fields and languages. These projects deal with biology (Dolbey et al., 2006), biomedicine (Verdaguer, 2020), football (Schmidt, 2009; *Dicionário da copa de mundo*, 2020), or Olympic games in general (Chishman et al., 2021), law (Pimentel, 2013), computing (Ghazzawi, 2016), food (Ortego-Antón, 2021), and linguistics (Malm et al., 2018). Other terminology projects, such as EcoLexicon (Faber et al., 2016) in the field of the environment, mention Frame Semantics without referring explicitly to the FrameNet methodology.
4 http://olst.ling.umontreal.ca/dicoadventure/ [Last accessed: 28/09/2023].
5 http://olst.ling.umontreal.ca/cgi-bin/dicoenviro/search_enviro.cgi [Last accessed: 28/09/2023].
6 http://olst.ling.umontreal.ca/dicoenviro/framed/index.php [Last accessed: 28/09/2023].
7 The Italian version of *DiCoEnviro* is developed with the collaboration of Maria Francesca Bonadonna of the University of Verona (Italy).

introduces a few basics about Frame Semantics and FrameNet concentrating on those components that are useful for the remainder of the chapter. Then, Section 4 describes how solutions were implemented in *DicoAdventure* and *DiCoEnviro* to answer the questions presented in Section 2. A methodological approach is summarised in Section 5. Finally, Section 6 sums up and mentions some future directions that resource compilation guided by Frame Semantics and FrameNet can take.

2. A Few Specific Questions

In this section, we examine the four questions listed at the beginning of this chapter from the perspectives of the *DicoAdventure* and *DiCoEnviro* resources.

Question 1: What kinds of terms should be taken into consideration?

When the *DicoAdventure* resource was designed and upon examining the content of a corpus assembled for the purpose of the project (i.e., the ADVENCOR corpus), it soon became quite obvious that important concepts in the field of adventure tourism were expressed linguistically as verbs. Verbal terms, such as *trek*, *journey*, *abseil*, *glide*, or *dive* expressing activities were used quite frequently in the corpus and stood out from the lists of candidate terms (Durán-Muñoz & L'Homme, 2020). This is hardly surprising since adventure tourism consists, after all, of various outdoor activities, but this observation contrasts with a tradition in domain-specific resources that usually focus on nouns. Although other parts of speech also needed to be analysed, it was decided early on that verbs would be the focus of the first steps of the compilation process. This also meant that terminologists involved in the *DicoAdventure* project would need to seek models that would best account for the linguistic properties of verbs (and other predicative units, such as nouns that express activities).

Similar questions arose when the *DiCoEnviro* project was launched. Important concepts linked to the environment (in topics such as climate change, endangered species, renewable energy) correspond to events, activities, and properties. Events and activities are expressed by verbs or nouns (*biodegrade*, *contaminate*, *impact* – noun or verb –, *reflect*, *storm*) and properties are expressed by adjectives or nouns (*biodiversity*, *endangered*, *green*, *sustainable*). Terminologists involved in this project also needed to turn to models that would help them account for these types of terms adequately. Methods in terminology are well suited to describe nouns (as single-word terms or as noun phrases), but say little about how verbs or adjectives should be handled.

Question 2: Depending on their nature, do terms behave differently?

As mentioned above, both the *DicoAdventure* and the *DiCoEnviro* projects needed to deal with terms that belong to parts of speech for which terminology methods have little to offer. Verbs, adjectives, and nouns that express activities, events, and properties are predicative units and as such require arguments, as shown in examples (1) and (2) taken from the ADVENCOR corpus and an environment corpus.

Examples in (1) show that the activity expressed by the verb *travel* is carried out by a first argument with the role TOURIST (realised with *we* and *many trekkers* in (1a) and (1b), respectively, and not instantiated in (1c) since the verb is used in the imperative form). A second argument expresses a location: in (1a) and (1b), the location is a path through which the first argument moves (*through the Brachina Gorge*; *through isolated areas*); in (1c), a source (*from San José*) and a destination (*to La Fortuna*) are realised. Examples in (2) show that the property expressed with *vulnerable* require two arguments: one is a cause (*to climate change*; *to economic damage from weather extremes*); the second one is an entity affected by this cause (*polar regions*; *modern societies*).

ADVENCOR: *travel*

(1) a. We [TOURIST] TRAVEL *the geological trail* [PATH] *through the Brachina Gorge* [PATH].
 b. *Visiting the most beautiful peaks* many trekkers [TOURIST] TRAVEL *through isolated areas* [PATH].
 c. TRAVEL *from San José* [SOURCE] *to La Fortuna* [DESTINATION], *and opt to swim at a local swimming hole at the base of a waterfall during free time in the afternoon.*

Environment corpus: *vulnerable*

(2) a. *Polar regions* [PATIENT] *are highly* VULNERABLE *to climate change* [CAUSE].
 b. *Modern societies* [PATIENT] *are becoming more* VULNERABLE *to economic damage from weather extremes* [CAUSE].

The approach usually taken to describe terms (in domain-specific resources) consists in establishing a connection between the linguistic unit or expression and a concept, that is, a generalisation of a group of objects of the real world. Definitions are formulated in such a way as to capture the clear-cut delimitation of concepts and their position in a conceptual structure. This approach has been used efficiently to account for nouns denoting entities, but cannot fully capture the meaning of other kinds of terms, especially those which require arguments, such as *travel*, *vulnerable*, and many others. Again, designers of *DicoAdventure* and *DiCoEnviro* needed to find alternatives: an adequate description of predicative terms should represent arguments since they are core components of the meaning of these terms.

Question 3: Which terms are related and how?

Taking into consideration relations between terms or between concepts expressed by terms is a fundamental part of terminology work regardless of the theoretical framework chosen or method applied. However, when various kinds of terms are considered (such as verbs and adjectives), new kinds of relations are uncovered, which differ from those that are taken into consideration in domain-specific resources (such as taxonomies, meronymies, or causal relations).

For instance, in the *DicoAdventure* and *DiCoEnviro* projects, the following relations involving predicative terms are quite common:

1. Terms that convey the same meaning but belong to different parts of speech. For example, the verb *travel* (*We* TRAVEL *the geological trail*) and the noun *travel* (*his extensive* TRAVELS *[...] from British Columbia to Norway to the White Salmon River*) express the same activity in adventure tourism. These two terms cannot be defined as true synonyms since they do not belong to the same part of speech and thus differ in terms of syntactic behaviour. Similarly, the adjective *vulnerable* and the noun *vulnerability* express the same property in the environment.
2. Series of terms that convey related meanings. For instance, in adventure tourism, the following verbs and nouns are all associated with the general idea of leaving home and going to another location for a given duration: *to travel, travel, adventure, trip, to tour, tour*. In the environment, *vulnerable, threatened,* and *vulnerability* all have something to do with the state of a given entity.
3. Predicative terms and their participants. For example, there is an obvious relation between *travel* and *traveller*, the latter being the person carrying out the activity expressed by the former; the same relation holds between *tour* and *tourist* in adventure tourism, and between *climatology* and *climatologist* in the environment.
4. Shared semantic components. In the environment, other sets of terms are linked to *vulnerable, threatened,* and *vulnerability*, which were mentioned in number 2. above, such as *harmful, risk, threat, imperil, threaten, endanger,* and so on. While *vulnerable, threatened,* and *vulnerability* profile the entity in a vulnerability situation (example (2)), *harmful* denotes a property which emphasises the cause (example (3)). *Imperil, threaten,* and *endanger* also emphasise the cause (example (4)) but denote an activity; *risk* and

threat emphasise a result or a consequence of the situation (example (5)). In the environment, other sets of terms sharing semantic components can be identified. For instance, *erode, erosion, eroded,* and *eroding* all share the semantic component of "change"; *pollute, pollution, polluted, polluting, polluter,* and *depollute* all share the semantic component of "damaged by substances that were introduced".

(3) a. *Such varied rates of migrations* [CAUSE] *can be particularly* HARMFUL *to species already under stress from climate and other factors* [PATIENT].
(4) a. *Increased sea-levels* [CAUSE] *could* IMPERIL *to species with limited distributions in coastal areas* [PATIENT].
   b. *Loss of important habitats (wetlands, tundra, isolated habitats)* [CAUSE] *would* THREATEN *some species, including rare/endemic species* [PATIENT].
   c. *Rising sea levels* [CAUSE] ENDANGER *coastal areas and small islands* [PATIENT].
(5) a. *Vulnerability refers to the* RISK *of adverse impacts* [RESULT] *from climate change, including extreme weather events and sea level rise, on both natural and human systems* [CAUSE].
   b. *A* THREAT *of extinction* [RESULT] *due to lack of possibility to migrate upward* [CAUSE].

Despite obvious connections, very few domain-specific resources record these relations (perhaps with the exception of the first one that would be considered to be a specific case of variation). Users of resources that record predicative terms would benefit from the representation of these relations. They could access a larger set of related terms and perhaps discover terms with which they are unfamiliar.

Question 4: Can a resource represent both knowledge structure and the linguistic properties of terms?

Terminologists often attempt to capture the knowledge structure of the domain they are asked to describe either before starting to collect terms and information about them or during the compilation process. In most terminology projects, it is assumed that there is a direct or indirect connection between terms and the knowledge they express and different methods can be used to represent these connections.

One example of a knowledge representation in a specialised field of knowledge is the domain ontology. Figure 1 illustrates how the concept "habitat" is represented in the Envo ontology,[8] an ontology of environment concepts.

---

8   https://www.ebi.ac.uk/ols/ontologies/envo [Last accessed: 28/09/2023].

"Habitat" appears in a taxonomy (which captures hierarchically organised "is-a" relations) and is linked to the concept "biosphere". "Habitat" inherits the properties assigned to superordinate concepts, including those assigned to the concepts to which the superordinate "biosphere" is linked. "Habitat" has specific properties that differentiate it from "biosphere" and other neighbouring concepts. The concept "habitat" can also be formally linked to other concepts via other relations. For instance, in Envo it is defined as a part of a "biome".

This simple example shows how an entity concept is represented in Envo. The relations established between "habitat" and other concepts are those that can be found between entities (such as "is-a" or "part-of"). The addition of activity, event, and property concepts to this formal representation is possible but raises different kinds of challenges, one of which consists in defining new sets of relations for concepts which are not entities. Another important challenge when using this kind of representation is accounting for the linguistic properties of terms. Often, given the abstract nature of ontologies, linguistic information is kept to a minimum (for instance, listing different linguistic labels, i.e., variants, in an annotation component, a much less formal part of the ontology).

Another example of knowledge representation was suggested for adventure tourism by Durán-Muñoz (2016) and is reproduced in Figure 2. This representation accounts for the general conceptual classes that are relevant in this domain. Interestingly, the representation provides for Action and Activity concepts and defines relations between these concepts and entity concepts labelled as Agent, Instrument, and Location. Although this second example considers Activity and Action concepts, it does not give a detailed account of more specific predicative units. We will see how this can be done in Section 4.1.

Figure 1. The concept "habitat" in the Envo ontology

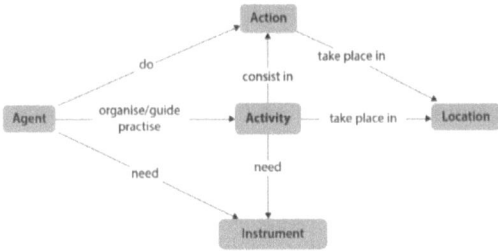

Figure 2. A general representation of general concept classes in adventure tourism (Durán-Muñoz, 2016)

3. Frame Semantics and FrameNet

Frame Semantics (Fillmore, 1976, 1982; Fillmore & Baker, 2010) and its methodological component as implemented in FrameNet (2023; Ruppenhofer et al., 2016) offer a certain number of principles and tools that can help terminologists tackle the questions discussed previously. This section summarises some of the basic assumptions and principles of Frame Semantics. Afterwards, we explain how these were applied to terminological data in the *DicoAdventure* and *DiCoEnviro* resources.

Frame Semantics is based on the assumption that the meanings of lexical units (henceforth, LUs) are constructed in relation to background knowledge (this knowledge is built on previous experience, on beliefs, and on social conventions) and that LUs are explained according to this background. For instance, a prototypical travelling situation is associated with the following background knowledge (most of which is shared by members of a community with similar experiences, beliefs, and social conventions): an activity carried out by people, going away from home, going to one or several destinations, or moving in an area, and so on. Interestingly, during the COVID-19 pandemic, the situation was strongly associated with the idea of being deprived of the possibility of carrying out this activity. Perhaps, this latter feature would not belong to a prototypical account of travelling.

In Frame Semantics, background knowledge associated with situations such as the travelling situation is captured in frames that represent cognitive and linguistic features: participants (obligatory and optional), props, nature of participants, and LUs that evoke the frame, among others. For instance, in FrameNet the situation described above is captured in a frame called TRAVEL. Participants in a situation, such as the people doing the travelling or the place where the travelling is done, are called frame elements (henceforth, FEs). Some FEs are compulsory (core FEs), since they are always cognitively active in a frame; others are optional (non-core FEs) and are activated only occasionally. In the TRAVEL frame, TRAVELER, AREA, GOAL, DIRECTION correspond to core FEs; DISTANCE, DURATION, and MANNER are defined as non-core FEs. As abovementioned, FrameNet also lists the LUs that evoke each frame. The following list is the one given for English in the TRAVEL frame:[9] *commute.v, excursion.n, expedition.n, getaway.n, jaunt.n, journey.n, journey.v, junket.n, odyssey.n, peregrination.n, pilgrimage.n, safari.n, tour.n, tour.v, travel.n, travel.v, traveler.n, trip.n, voyage.n, voyage.v.*

The frame that we just summarised is accompanied by linguistic descriptions enabling a connection between the conceptual characterisation of a situation and the linguistic units or expressions used to convey it. The first linguistic module is lexical, and it provides additional information about LUs. In each entry, definitions of the LUs are given. For instance, the verb *journey*, when it evokes the TRAVEL frame, is defined as "make a journey for pleasure to several locations" (FrameNet, 2023). Lexical items can evoke different frames if they are polysemous. Syntactic information is also provided to show how FEs combine with this LU in corpora.

A second linguistic module presents sentence annotations for each LU. The LU that evokes a frame and the FEs that appear in the sentences are annotated; thus, the label used to represent them (TRAVELER, AREA, GOAL, DIRECTION, and so on) is reproduced and syntactic information is added. Some non-instantiated FEs (for instance, the first FE of a transitive verb used in the passive can be omitted) are mentioned in the annotations. Figure 3 shows a sample of contextual annotations for the verb *tour* when it evokes the TRAVEL frame.

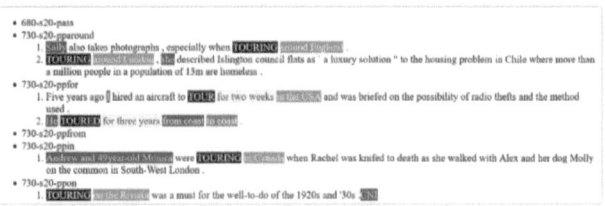

Figure 3. Annotations of example sentences for the verb *tour* in FrameNet

---

9   https://framenet2.icsi.berkeley.edu/fnReports/data/frameIndex.xml?frame=Travel [Last accessed: 28/09/2023].

Finally, frames can be connected to provide a broader perspective on situations and how they are linked to other ones. Figure 4 shows the different relations shared by the TRAVEL frame with other frames. It inherits from the SELF_MOTION frame (evoked by LUs such as *climb*, *run*, *venture*, and *walk*) and has a subframe called SETTING_OUT (evoked by LUs such as *decamp*, *set off*, and *set out*).

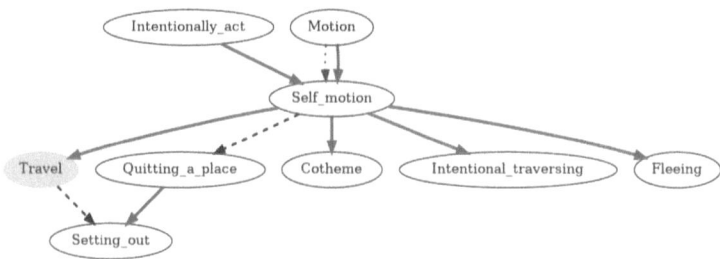

Figure 4. Relations between TRAVEL and other frames in the FrameGrapher

4. Solutions in *DicoAdventure* and *DiCoEnviro*

There are different ways to answer the questions listed in Sections 1 and 2 depending on the theoretical framework chosen, the methodology derived from it, and the specific objectives a domain-specific resource seeks to meet. This section examines solutions inspired by Frame Semantics and FrameNet and, more specifically, how these solutions were implemented in *DicoAdventure* and *DiCoEnviro*.

Section 4.1. shows how answers to questions 1 and 2 were implemented in both resources. Since the *DiCoEnviro* project started over a decade ago, implementations of solutions to questions 3 and 4 are examined in this resource only (Sections 4.2. and 4.3.). However, we believe that similar strategies could be applied to *DicoAdventure*. Mention should be made that, even if both resources apply the general methodological principles developed in FrameNet, a few adaptations were made and are mentioned whenever relevant.

4.1. Questions 1 and 2: Different kinds of terms and different linguistic behaviours

As mentioned in Section 1, *DiCoEnviro* records terms that belong to different parts of speech, and thus includes predicative terms. It was decided early on by the members of the project that the argument structure would be a central data category in the entries for predicative terms. The arguments are identified using the contexts extracted from the corpora and, in the current version of *DiCoEnviro*, they appear in an argument structure which states their number and position according to the predicative unit. Additionally, each argument is labelled using two different systems: (1) a typical term (which should be representative of the kinds of terms that can be used in that position), and (2) a semantic role label (that is less transparent for users, but which is used throughout the entry for consistency).

The entry for the adjective *vulnerable* is reproduced in Figure 5. The argument structure accounts for two different arguments: the first one is labelled as a patient; the second one, as a cause. Typical terms chosen are *ecosystem* and *population* for the patient, and *change* for the cause. Users can also view other possible realisations of arguments by clicking on the + sign placed next to the typical terms.

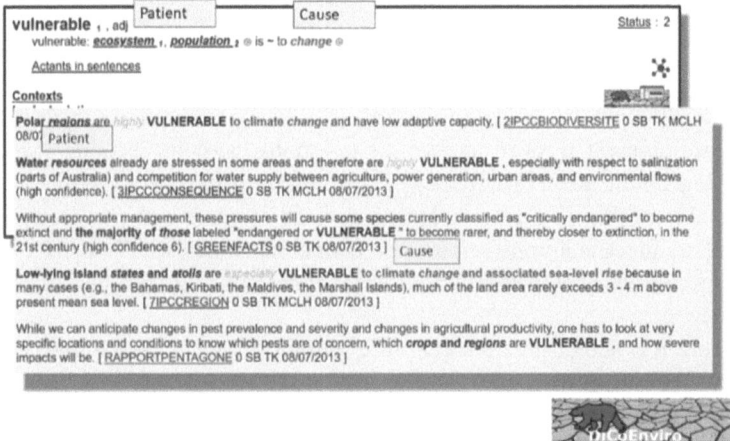

Figure 5. Part of the entry for *vulnerable* in *DiCoEnviro*

In addition to the statement of the argument structure, up to 20 contexts are annotated applying the method developed in FrameNet (Ruppenhofer et al., 2016). All contexts are extracted from specialised texts (with a hyperlink to view the reference). In contexts, core FEs (or arguments) are identified and labelled with their semantic role. In annotations, non-core FEs (or circumstantials) are also annotated. Additionally, a summary is included to give a picture of the syntactic groups and functions of participants found in the annotations.

In contrast to FrameNet, the approach taken in *DiCoEnviro* is essentially bottom-up: the first steps of the compilation process consist in identifying terms and describing their linguistic properties. Hence, instead of defining a frame with FEs at a conceptual level, terminologists of the *DiCoEnviro* project make a selection of terms based on the content of specialised corpora and then describe their argument structure. The relationship of terms with conceptual situations is done at a later stage (this is explained in Section 4.3.). Furthermore, in *DiCoEnviro* reference is made to arguments, instead of core FEs, and to circumstantials, instead of non-core FEs. The labels given to participants also differ from those used in FrameNet. A more widely applicable set of labels is used, whereas in FrameNet many labels are defined within specific frames.

The current version of *DicoAdventure* records verbs and some nouns that denote activities and, as in *DiCoEnviro*, a decision was made early on to represent the argument structure explicitly. Up to now, verbs recorded in the resource are motion verbs which can display multiple arguments (among other linguistic specificities). Figure 6 shows the first two sections of the entry for the verb *travel₁*. The argument structure states the number of possible arguments and their position according to the term. As in *DiCoEnviro*, labels used for arguments differ from those used in FrameNet; in fact, some of the labels are specific to the domain of adventure tourism (TOURIST, VEHICLE, etc.) (Durán-Muñoz & L'Homme, 2020). Additionally, a table placed under the argument structure lists the different realisations of these arguments in the corpus; users can select a labelled argument and visualise the realisations (as shown in Figure 6 for DESTINATION).

Figure 6. Extract from the entry for *travel₁* in *DicoAdventure*

In *DicoAdventure*, each entry comes with over 10 contexts which are annotated in a way that is reminiscent of the method used in *DiCoEnviro* and FrameNet (cf. Figure 7).

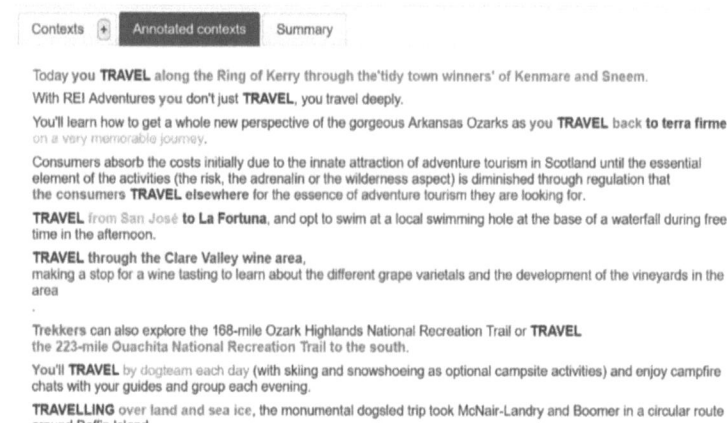

Figure 7. Some annotations for *travel*₁ in *DicoAdventure*

4.2. Question 3: Locating semantically related terms

As mentioned in Section 2, taking predicative terms into consideration uncovers sets of relations that domain-specific resources seldom describe, for example, terms that carry the same meaning but belong to different parts of speech, terms that share close meanings without being true synonyms, terms that share semantic components, and so forth.

Let us consider the examples in (6) taken from an environment corpus:

(6) a. *Polar regions* [PATIENT] *are highly* VULNERABLE *to climate change* [CAUSE].
 b. *The number of species* [PATIENT] THREATENED *by extinction* [CAUSE].
 c. *Paths that generate lesser stresses are expected to lessen the* VULNERABILITY *of human and natural systems* [PATIENT] *to climate change* [CAUSE].
 d. *The* SENSITIVITY *of major cereal and tree crops* [PATIENT] *to changes in temperature, moisture and CO2 concentration of the magnitudes projected for the region* [CAUSE] *has been demonstrated in many studies.*

In environment texts, *vulnerable, threatened, vulnerability,* and *sensitivity* express the property of an entity to be susceptible to change (usually in a negative way) due to an external cause. Although the meaning of these four terms is related, they cannot be defined as true synonyms, since they are not interchangeable in all contexts. Furthermore, two of the terms listed are adjectives and the other two are nouns, which rules out synonymy. These terms share other similarities: they all have two arguments of a similar nature (a patient and a cause) and share some circumstantials, such as units or phrases expressing a degree (*highly, more, much more, especially; increased, high*), and they profile the entity being subjected to a cause. Although we are not dealing with true synonyms, there is enough linguistic evidence to indicate that these terms evoke the same situation in the environment.

Series of terms that are recorded in *DiCoEnviro* and are found to evoke the same situation are grouped into frames. This was done for the four terms in example (6). Frames are represented in *Framed DiCoEnviro*, a resource which is superimposed on *DiCoEnviro*. Table 1 shows how the terms listed above are placed in a frame called BEING_AT_RISK, based on the frame recorded in FrameNet. Interestingly, terms in other languages are also listed when they evoke the same frame.

Table 1. The BEING_AT_RISK frame in *DiCoEnviro*

BEING_AT_RISK		
Definition: A Patient is in a state where it is exposed to or otherwise liable to be affected by a Cause.		
Example(s): [EN] *Species such as walrus, seals, and polar bears will be THREATENED.* (Source: 3IPCCCONSEQUENCE)		
[EN] *The VULNERABILITY of human societies and natural systems to climate extremes is demonstrated by the damage, hardship, and death caused by events such as droughts, floods, heat waves, avalanches, and windstorms.* (Source: GREENFACTS)		
[FR] *L'ALENA indique qu'en Amérique du Nord seulement, 235 espèces sauvages sont MENACÉES d'extinction .* (Source: 3CANADAENVIRON)		
Notes: This frame is based on BEING_AT_RISK in FrameNet Click here to see associated FrameNet infos		
Participants (1): 1. Patient 2. Cause	Participants (2): 1. Degree (14) 2. Descriptor (2) 3. Cause (1)	
English LUs: • sensitivity$_1$ • threatened$_1$ • vulnerability$_1$ • vulnerable$_1$	French LUs: • menacé$_1$ • sensible$_1$ • vulnérable$_1$	Spanish LUs: • amenazado$_1$

A frame layer has not been added to *DicoAdventure* yet. Nevertheless, many entries already recorded share the kinds of similarities that were mentioned for *threatened, sensitivity, vulnerable*, and *vulnerability* and are likely to evoke specific situations. For instance, examples in (7) show that *travel* (verb and noun), *expedition*, and *tour* (verb and noun) share arguments and circumstantials (Tourist, Source, Destination, Place, etc.). Other verbs and nouns found in the Advencor corpus can also be linked to the same arguments and circumstantials: *trip, to journey, a journey*, among others. This evidence could indicate that these terms evoke a situation similar to the TRAVEL situation described in Section 3 about FrameNet with some features that would be specific to adventure tourism. The series of terms could also be captured in a frame or with another representation model that would make their connections to a given situation explicit for users of *DicoAdventure*.

(7) a. Travel *from Las Vegas* [Source] *to Bryce Canyon* [Destination].
 b. *His* [Tourist] extensive travels *and kayaking adventures on diverse bodies of water* [Place] *from British Columbia* [Source] *to Norway* [Destination].
 c. Expeditions *to the world's highest peaks* [Destination].
 d. *Next, on Prijevor, we* [Tourist] tour *the mountain "Katuns"* [Path].
 e. *Take a 30-minute scenic helicopter* tour *over the Great Barrier Reef* [Place] *from Cairns* [Source].

4.3. Question 4: Connecting linguistic properties to knowledge

The last question addressed in this chapter is a difficult one, but remains of utmost importance for designers of domain-specific resources: how can a resource account for the linguistic properties of terms while taking the knowledge structure of domains into consideration? It seems difficult to reconcile these two objectives in the same resource in a way that remains consistent. A choice must usually be made by designers as to the approach to be taken; each emphasises one aspect and overlooks or minimises the other (Hirst, 2009; L'Homme, 2018; L'Homme & Bernier-Colborne, 2012).

Frame Semantics and FrameNet can offer a solution to this thorny issue, since an abstract conceptual representation (a frame) is connected to the linguistic properties of terms. Section 4.2. explained how frames could be discovered based on shared linguistic properties. We can take this one step forward and try to find, among frames described in a given field of knowledge, which ones are connected to others. This strategy is also based on FrameNet (cf. Section 3, Figure 3), but is adapted to specific areas of knowledge. Once identified, these connections can start saying something about the contribution of different kinds of terms to the knowledge structure of a domain.

For instance, in *Framed DiCoEnviro*, the BEING_AT_RISK frame (which the terms *threatened, sensitivity, vulnerable,* and *vulnerability* evoke) appears in a broader scenario called 'Risks and ways to prevent them' (Figure 8). It offers a perspective on[10] the RISK_SCENARIO frame which differs from the one given by the RISKY_SITUATION frame. In this case, the terms provide a different point of view on a situation: the terms in the BEING_AT_RISK frame place the entity at risk in the foreground and the cause in the background; in the RISKY_SITUATION frame, the point of view is reversed since the cause is placed in the foreground and the entity, in the background. RISKY_SITUATION is evoked by terms such as *endangerment, harmful, hazard,* and *hazardous*.

Other frames are also linked to the BEING_AT_RISK frame. The RUN_RISK situation is evoked by terms such as *risk* and *threat* and emphasises a result or consequence. The frame ENDANGERING is linked to the BEING_AT_RISK frame through a "is causative of" relation and is evoked by terms such as *endanger, imperil,* and *threaten*.

In *Framed DiCoEnviro*, different scenarios were identified in this way ("Changes affecting the environment", "Species activities and life", "Management of waste", etc.). These scenarios, which are all defined on the basis of the linguistic properties of terms initially described in *DiCoEnviro*, can tell short stories about the environment. They are also enriched on a regular basis as more terms and frames are added to the resources. Ultimately, they allow us to obtain a picture of the contribution of different kinds of concepts to the conveyance of knowledge in a domain.

This work has not been carried out yet in the *DicoAdventure* resource, but it is likely that similar scenarios could be developed once more data is encoded in it and frames are defined on this basis.

---

10 The perspective on a relation is based on the one defined in FrameNet (Ruppenhofer et al., 2016).

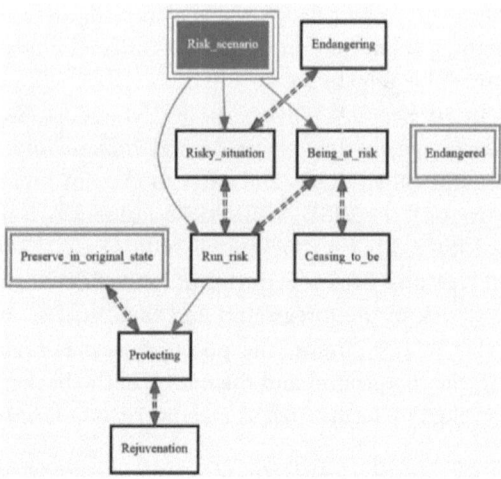

Figure 8. The 'Risks and ways to prevent them' scenario in *Framed DiCoEnviro*

5. Integrating this Approach into a Methodology

This short section gives a brief overview of the methodological steps that are taken in the *DiCoEnviro* project to compile both *DiCoEnviro* and *Framed DiCoEnviro*. It shows more concretely how the different questions raised in this chapter are addressed when handling the data.[11] In the *DicoAdventure* project, a similar methodology was set up and covers more or less the first five steps described in this section (Durán-Muñoz & L'Homme, 2020). The methodological frameworks are quite similar with some differences here and there, the main one being that, in *DicoAdventure*, the focus is placed on verbal terms during the first stages of the project.

These first five steps lead to the compilation of terminological entries in *DiCoEnviro* and implement solutions given to questions 1 (different types of terms) and 2 (linguistic behaviour of terms):

1. Corpora on specific topics linked to the environment are compiled by applying explicit criteria to select texts.
2. Terms are then identified with a method that combines term extraction (Drouin, 2003) and manual filtering of the results. At this stage, single-word

---

11 More specific details can be found in L'Homme et al. (2020).

terms that belong to the parts of speech of noun, verb, adjective, and adverb are considered. Manual filtering applies lexico-semantic criteria to validate terminological status. At this stage and the following one, different meanings conveyed by polysemous units must be differentiated. Meanings that are not linked to the environment are set aside since they are not recorded in *DiCoEnviro*.
3. For each validated term, up to 20 contexts (to be annotated at a later stage) are extracted from the corpus.
4. A first version of the argument structure is formulated. It is validated with contextual annotations.
5. The selected contexts are annotated.

Once a certain number of terms linked to the same environmental topic are recorded in *DiCoEnviro* and the contexts in which they appear are annotated, the following three steps can be carried out. This process leads to the enrichment of *Framed DiCoEnviro*. They address questions 3 (finding related terms) and 4 (connecting terms and knowledge structure):

6. Frames are defined based on the observation of similarities between terms: argument structure, shared circumstantials, denotation of similar situations, and so forth. At this stage, the FrameNet is consulted to see if similar frames have been encoded.
7. Once frames are established based on the data in *DiCoEnviro*, they are encoded as in Table 1.
8. Finally, links between frames are established using basically the set of relations defined in FrameNet (e.g., "Inheritance", "Causative", "Subframe").

Users can access *Framed DiCoEnviro* from the terminological entries in *DiCoEnviro* and vice versa. Links to FrameNet are also provided when frames are based on the original FrameNet resource.

6. Conclusions

During the past decades, terminologists have turned to alternative frameworks (corpus linguistics, discourse analysis, cognitive linguistics, lexical semantics, etc.) to tackle problems for which more traditional approaches to terminology provide few solutions. Innovative approaches and resources are now developed to describe terms and represent some of their properties. This chapter showed how principles borrowed from Frame Semantics provide helpful answers to questions raised by domain-specific resource compilation and examined how these principles are implemented concretely in two resources.

Using specific examples taken from the fields of adventure tourism and the environment, we showed how resources can account for different kinds of terms, more specifically, predicative terms. Enhanced descriptions of argument structures (predicative terms and their participants) can be implemented; annotated contexts provide information about the behaviour of terms in specialised corpora; syntactic features can be derived from annotations. Frame Semantics and FrameNet are also helpful to capture relations that are seldom accounted for in terminological resources and allow, more specifically, to identify and represent terms that evoke situations in specialised domains. The framework can then be extended to the identification of larger scenarios based on frames that are connected to each other. This can be considered to be a kind of knowledge structure, albeit different from the ones currently used to represent concept structures in terminology.

The *DiCoEnviro* and the *DicoAdventure* are both ongoing projects, and entries are added in the corresponding resources in different languages on a regular basis; also new frames are defined in *Framed DiCoEnviro*. The methodology is stable and can be extended to new languages or new areas of knowledge. In *DiCoEnviro*, the method has been applied to Chinese, English, French, Portuguese, and Spanish terms extracted from different environment corpora (endangered species, renewable energy, transportation electrification, etc.). In *DicoAdventure*, it has been applied to English and Spanish motion verbs and nouns related to adventure activities. These resources differ quite sharply from other terminological repositories that account for terms or specialised concepts (e.g., Term Banks –such as IATE[12] or Termium Plus®[13]–, thesauri – such as GEMET[14]–, or ontologies –such as the Envo ontology; cf. Section 2–), and they are viewed as alternative models to obtain a different perspective on terminological data.

The chapter examined a short list of questions, and much more could be said about the benefits of applying a Frame Semantics framework to the description of terms. Frame Semantics can provide some interesting insights into specialised fine-grained polysemy (L'Homme, 2021, 2023), and it can also be used to highlight equivalence degrees (Pimentel, 2013). Given that the framework was applied originally to the description of everyday situations, studies on the similarities and differences between the general lexicon and the terminology of a

---

12 https://iate.europa.eu/home [Last accessed: 28/09/2023].
13 https://www.btb.termiumplus.gc.ca/ [Last accessed: 28/09/2023].
14 https://www.eionet.europa.eu/gemet/en/themes/ [Last accessed: 28/09/2023].

domain or across domains would also highlight some similarities and differences between the conceptualisations of situations.

References

Bergenholtz, H., & Tarp, S. (1995). *Manual of specialised lexicography*. John Benjamins Publishing Company.

Chishman, R., da Silva, B., Nardes dos Santos, A., Vianna, A. L. T., de Oliveira, S., Martins, M., & de Schryver, G.-M. (2021). Building a paralympic, frame-based dictionary – Towards an inclusive design for *Dicionário Paraolímpico* (Unisinos/Brazil). In L. Mitits & S. Kiosses (Eds.), *Lexicography for Inclusion: Proceedings of the 19th EURALEX International Congress, 7–9 September 2021, Alexandroupolis, Vol. 2* (pp. 723–731). Democritus University of Thrace.

Dolbey, A., Ellsworth, M., & Scheffczyk, J. (2006). BioFrameNet: A domain-specific FrameNet extension with links to biomedical ontologies. In O. Bodenreider (Ed.), *Proceedings of the Second International Workshop on Formal Biomedical Knowledge Representation: "Biomedical Ontology in Action" (KR-MED 2006)* (pp. 87–94). National Library of Medicine.

Drouin, P. (2003). Term extraction using non-technical corpora as a point of leverage. *Terminology, 9*(1), 99–117. https://doi.org/10.1075/term.9.1.06dro

Durán-Muñoz, I. (2016). Producing frame-based definitions: A case study. *Terminology, 22*(2), 223–249. https://doi.org/10.1075/term.22.2.04mun

Durán-Muñoz, I., & L'Homme, M.-C. (2020). Diving into English motion verbs from a lexico-semantic approach: A corpus-based analysis on adventure tourism. *Terminology, 26*(1), 34–60. https://doi.org/10.1075/term.00041.dur

Faber, P., León-Araúz, P., & Reimerink, A. (2016). EcoLexicon: New features and challenges. In I. Kernerman, I. Kosem Trojina, S. Krek, & L. Trap-Jensen (Eds.), *GLOBALEX 2016: Lexicographic Resources for Human Language Technology in conjunction with the 10th edition of the Language Resources and Evaluation Conference* (pp. 73–80). http://www.lrec-conf.org/proceedings/lrec2016/workshops/LREC2016Workshop-GLOBALEX_Proceedings-v2.pdf

Fillmore, C. J. (1976). Frame Semantics and the nature of language. *Annals of the New York Academy of Sciences, 280*(1), 20–32. https://doi.org/10.1111/j.1749-6632.1976.tb25467.x

Fillmore, C. J. (1982). Frame Semantics. In The Linguistic Society of Korea (Ed.), *Linguistics in the morning calm* (pp. 111–137). Hanshin.

Fillmore, C. J., & Baker C. (2010). A frames approach to semantic analysis. In B. Heine & H. Narrog (Eds.), *The Oxford handbook of linguistic analysis* (pp. 313–340). Oxford University Press. https://doi.org/10.1093/oxfordhb/9780199544004.013.0013

FrameNet Brasil. (2020). *Dicionário da copa de mundo.* https://www2.ufjf.br/framenetbr/dicionario/

Ghazzawi, N. (2016). *Du terme prédicatif au cadre sémantique: méthodologie de compilation d'une ressource terminologique pour des termes prédicatifs arabes en informatique* [Unpublished doctoral dissertation, University of Montreal].

Hirst, G. (2009). Ontology and the lexicon. In S. Staab & R. Studer (Eds.), *Handbook on ontologies* (pp. 269–292). Springer.

Lamberti Arraes, L. (2022). Descobrindo o léxico especializado no português brasileiro: Uma abordagem léxico-semântica. *Domínios de lingu@gem, 16*(3), 1146–1172. https://doi.org/10.14393/DL51-v16n3a2022-9

L'Homme, M.-C. (2018). Maintaining the balance between knowledge and the lexicon in terminology: A methodology based on Frame Semantics. *Lexicography, ASIALEX, 4,* 3–21. https://doi.org/10.1007/s40607-018-0034-1

L'Homme, M.-C. (2021). Revisiting polysemy in terminology. In Z. Gavriilidou, M. Mitsiaki, & A. Fliatouras (Eds.). *Lexicography for Inclusion: Proceedings of the 19th EURALEX International Congress, 7–9 September 2021, Alexandroupolis, Vol. 1* (pp. 415–424). Democritus University of Thrace.

L'Homme, M.-C. (2023). Managing polysemy in terminological resources. *Terminology* (online). https://doi.org/10.1075/term.22017.lho

L'Homme, M.-C., & Bernier-Colborne, G. (2012). Terms as labels for concepts, terms as lexical units: A comparative analysis in ontologies and specialized dictionaries. *Applied Ontology, 7*(4), 387–400. https://doi.org/10.3233/AO-2012-0116

L'Homme, M.-C., Robichaud, B., & Subirats, C. (2020). Building multilingual specialized resources based on FrameNet: Application to the field of the environment. In T. Torrent, C. F. Baker, O. Czulo, K. Ohara, & M. R. L. Petruck (Eds.), *Proceedings of the International FrameNet Workshop 2020: Towards a Global, Multilingual FrameNet* (pp. 94–102). European Language Resources Association.

Malm, P., Mumtaz, S., Virk, L., Borin, L., & Savera, A. (2018). LingFN. Towards a FrameNet for the linguistics domain. In T. Timponi Torrent, L. Borin, & C. Baker (Eds.), *Proceedings of the LREC 2018 Workshop International FrameNet Workshop 2018: Multilingual Framenets and Constructicons* (pp. 37–43). ELRA.

Ortego-Antón, T. (2021). e-DriME. A Spanish-English frame-based dictionary about dried meats. *Terminology, 27*(2), 294–321. https://doi.org/10.1075/TERM.20013.ORT

Pimentel, J. (2013). Methodological bases for assigning terminological equivalents. A contribution. *Terminology*, *19*(2), 237–257. https://doi.org/10.1075/term.19.2.04pim

Ruppenhofer, J., Ellsworth, M., Petruck, M., Johnson, C., Baker, C., & Scheffczyk, J. (2016). *FrameNet II: Extended Theory and Practice.* https://framenet.icsi.berkeley.edu/fndrupal/index.php?q=the_book

Schmidt, T. (2009). The Kicktionary – A multilingual lexical resource of football language. In H. C. Boas (Ed.), *Multilingual FrameNets in computational lexicography. Methods and applications* (pp. 101–134). Mouton de Gruyter.

Verdaguer, I. (2020). Semantic frames and semantic networks in the Health Science Corpus. *Estudios de Lingüística del Español*, 117–155. https://doi.org/10.36950/elies.2020.8542

Zheng, Y. (2021). *Frame Semantics for the field of climate change: Discovering frames based on Chinese and English terms* [Unpublished Master's dissertation, University of Montreal].

Elena Manca

University of Salento
elena.manca@unisalento.it

8 Patterns and perspectives in the language of Italian and British walking holidays

Abstract: This chapter aims to contribute to the inter- and transdisciplinary research on adventure travel tourism by analysing the discursive patterns used in websites promoting walking holidays in Italy and in the UK. The analysis focuses on the investigation of the lexical, semantic, and pragmatic choices which characterise this type of discourse. More specifically, it aims to discuss how the interaction between authors and readers is achieved and how space is described in relation to participants. To reach this goal, a comparable corpus consisting of Italian and British websites advertising walking holidays was assembled and investigated through analytical methods typical of Corpus Linguistics. Discursive patterns were interpreted following the theoretical models of metadiscourse (Hyland, 2005) and spatial description (Levinson, 1996; Taylor & Tversky, 1996). Results shed light on the features of the promotion of Italian and British walking holidays with important applications in the marketing domain.

Keywords: corpora, metadiscourse, phraseology, spatial description, tourism discourse

1. Introduction

The effects of climate change will likely influence the future of tourism in the next decade. In particular, two significant challenges will have to be faced by the tourism sector: the transition to a low-carbon economy and the adaptation to the environmental and socio-economic impacts of a climate-disrupted world (Scott, 2021, p. 7). In this highly complex scenario, some forms of tourism may reveal more sustainable than others and, for this reason, they should be the object of inter- and transdisciplinary research. Adventure travel is based on physical activity in a natural environment and on the discovery of local cultures and folklore. The Adventure Travel Trade Association (ATTA)[1] includes among its aims and actions a series of educational programming and research to support

---

1 https://www.adventuretravel.biz/ [Last accessed: 28/09/2023].

the necessary transformations in travel in order to reduce emissions and help raise awareness among travellers.

The aim of this chapter is to contribute to the debate on this form of tourism by analysing the discursive patterns used on websites promoting walking holidays in Italy and in the UK. The analysis focuses on the investigation of the lexical, semantic, and pragmatic choices which characterise this type of discourse. More specifically, it aims to discuss how the interaction between authors and readers is achieved and how space is described in relation to participants. The study starts with the identification of the net of co-occurrences (Sinclair, 1996) of some frequent pronouns in a corpus representative of the discourse of walking holidays advertised on British and Italian websites. Next, collocations and patterns are interpreted by considering Hyland's (2005) interactional dimension of metadiscourse and theories on spatial description (Levinson, 1996; Taylor & Tversky, 1996). Results are discussed to outline some of the features of the British and Italian promotions of walking holidays, thus providing the basis upon which effective models of promotion can be developed.

The chapter is organised as follows: Section 2 and its subsections briefly outline the three theoretical models adopted for the analysis. Section 3 presents the material and methodology used, while Section 4 gives a detailed description of the analysis and the results achieved. Finally, Section 5 draws some conclusions and provides some indications for further applications of this study.

2. Perspectives: Repeated Linguistic Events, Metadiscourse, Spatial Description

This chapter addresses the notion of "perspective" by drawing on three theoretical frameworks: (1) the phraseological approach to language and the identification of repeated linguistic events through specifically designed corpora (Sinclair, 1996), (2) metadiscourse and the interactional linguistic strategies used to express authorial stance and to encode readers' engagement (Hyland, 2005), and (3) spatial description and the linguistic features of route perspectives (Levinson, 1996; Taylor & Tversky, 1996). The reason for combining these three theoretical frameworks is to be found in the complex interplay of voices and events in the description and promotion of walking holidays on websites. By applying a methodology based on Corpus Linguistics to identify regularities in language (Sinclair, 1991, 1996), functionally specialised usages of linguistic patterns on walking holidays' websites can be identified and described. The theoretical framework of metadiscourse allows us to understand how the websites' communicative aims are linguistically encoded and achieved and which strategies are used by holiday organisers to make themselves visible in a text and to address potential walkers. In addition, identifying the linguistic features used in spatial description provides insights into the strategies of tour description and how participants and elements of the surrounding environment relate to each other in a coherent and cohesive text. In the following sections, the three theoretical frameworks are briefly outlined.

2.1. Phraseological approach to language

The analysis described in this chapter starts from the observation of repeated linguistic events to map frequent lexical and semantic choices associated with pronouns and to identify the pragmatic meanings associated with these choices in the discourse of "walking holidays" tourism.

The phenomenon of collocation, that is to say, the frequent co-occurrence of words, shows that "words enter into meaningful relations with other words around them" (Sinclair, 1996, p. 71) and make meanings by their combinations. These "meaningful relations" between items and their environment imply that words cannot be selected independently in that lexical constraints operate at the level of word choice. The effects of these constraints are visible in the occurrence of repeated multi-word combinations, which constitute those 'semi-preconstructed phrases' (Sinclair, 1991, p. 110) that language users have available to them. This phraseological tendency of language is described in Sinclair's (1991, p. 110) Idiom Principle and contrasts with the traditional view of language, which sees text production as a series of free lexical choices. Repeated linguistic events are systematic; they can be counted and analysed, thus providing valuable insights into the strong relationship between words, texts, and culture. Manca (2004) describes this strong interdependence by comparing the collocational ranges of the Italian word *natura* and its English literal translation equivalent *nature*. Although the two words are provided as equivalent by dictionaries, an analysis of the two items in a comparable corpus of British and Italian tourist websites reveals that their patterns of usage are different. These differences reflect two different cultural approaches to nature: in Italian texts, *natura* is seen as a positive and unspoilt entity in which people immerse themselves to be purified; in British texts, *nature* is mainly used as an adjective to characterise the setting where activities are carried out.

2.2. Metadiscourse

Metadiscourse focuses on the way writers and speakers interact with readers and listeners through language. This interaction is encoded through a set of linguistic strategies that can be interactive or interactional (Hyland, 2005, p. 49). Interactive strategies are used to organise the contents in a text structurally. They include linguistic elements whose function is expressing relations, referring to discourse acts, sequences, and stages, relating to information in and out of the text, and elaborating propositional meaning. For this reason, the interactive dimension includes the following devices: (1) transition markers, (2) frame markers, (3) endophoric markers, (4) evidentials, and (5) code glosses. Conversely, interactional strategies group those items which are used by authors to make their perspective clear and explicit and to involve and engage their readers. This dimension includes five broad categories: (1) hedges, (2) boosters, (3) attitude markers, (4) self-mention, and (5) engagement markers.

Regarding the interactional strategies, the category of hedges includes those devices that indicate the subjectivity of a position and the degree of its negotiation; examples are *among the others*, *possible*, *might*, and *perhaps*. Boosters are used to close down possibilities of negotiation and emphasise certainty; some examples are adverbs such as *clearly* and *obviously*. Attitude markers indicate the writer's affective attitude to propositions and include attitude verbs (*agree*, *prefer*), sentence adverbs (*unfortunately*, *hopefully*), or adjectives (*appropriate*, *logical*, *remarkable*). Self-mention refers to the writers' presence in a text and describes how they stand in relation to their arguments and their readers; this category is linguistically expressed by first-person pronouns and possessive adjectives (*I*, *me*, *mine*, exclusive *we*, *our*, *ours*). On the contrary, engagement markers focus on readers and are used to attract their attention and involve them as participants. The linguistic options available to writers to perform both these functions are reader pronouns (e.g., *you*, *your*, inclusive *we*), interjections (e.g., *by the way*, *you may notice*), questions, directives (e.g., imperatives and obligation modals), and references to shared knowledge.

Since one of the aims of this chapter is to describe the perspective adopted by walking holiday organisers to address their potential customers, data from the English and Italian websites are interpreted based on the interactional dimension of metadiscourse, with a particular focus on hedges, boosters, self-mention, and engagement markers.

2.3. Spatial descriptions

Spatial descriptions linguistically locate objects with respect to a reference frame (Taylor & Tversky, 1996, p. 372). Scholars have identified three reference frames and three types of perspectives for spatial descriptions (cf. Buhler, 1982; Levelt, 1984, 1989; Levinson, 1996; Taylor & Tversky, 1996, among others). Levinson

(1996) develops a tripartite model of reference frames including the relative reference frame, the intrinsic reference frame, and the extrinsic or absolute reference frame. In the relative reference frame, coordinates are selected in relation to a speaker or to a viewer/addressee. The location of things is described with reference to the individual's front, back, left, and right, and with respect to another object in the scene (Levinson, 1996, p. 365). The intrinsic reference frame uses a specific object as the origin of the coordinate system, and things are located according to the object's intrinsic front, back, left, right, top, and bottom (Levinson, 1996, p. 365; Taylor & Tversky, 1996, p. 374). The extrinsic or absolute reference frame, as its name suggests, uses a coordinate system external to the scene and is based on cardinal directions (north, south, west, east). Building on these three reference frames, Taylor and Tversky (1996, pp. 376–377) elaborate three perspectives to explain how an environment can be experienced and described:

a. The gaze perspective, which uses a relative frame of reference and takes a fixed outside point of view.
b. The route perspective, based on the intrinsic frame of reference and describing the environment from a changing viewpoint.
c. The survey perspective, which takes a view from above and corresponds to the extrinsic reference frame.

In a study that examines the language used to describe environments and structural determinants of perspective choice, Taylor and Tversky (1996) show that the route and the survey perspectives are frequently combined. In a survey perspective (Taylor & Tversky, 1996, p. 385), landmarks are described relative to one another in a bird's eye view using the canonical direction terms *north*, *south*, *east*, and *west*. In addition, verbs are more stative than active, and frequent use of the verb *to be* can be observed. Fictive motion verbs (Talmy, 2000), such as *run*, *border*, *cross*, *turn*, and others are used to describe the path of a road or the course of a river. Conversely, in the route perspective (Taylor & Tversky, 1996, p. 375), speakers address listeners as *you*, taking a changing viewpoint from within the environment, and describe landmarks with respect to the listeners' imagined position. Therefore, in a route tour, subjects are the addressees and verbs are actions. Furthermore, according to Tversky (2004, p. 382), a route perspective describes at least two events linked in time while a survey perspective has declarations and not events, thus implying only the presence of spatial links. Taylor and Tversky (1996, p. 377) also suggest that the type of route description provided in response to a request for directions differs from the route descriptions given in response to a request for description. In directions, speakers tend to use imperatives and to mention only the essential landmarks, whereas in the case of descriptions, imperatives are not used, and more landmarks are mentioned. The two scholars also notice that descriptions, in order to be coherent, follow the given/new principle. Known spatial locations are, therefore, expected to be described first, while new landmarks are described in relation to them.

As noticed by Fulga (2012, p. 27), although space has universal physical properties, the location of people and objects in an environment is linguistically encoded differently by different languages or cultures. The same can be said for the encoding of movement (Fulga, 2012, p. 28). Indeed, the Source-Path-Goal schema (Lakoff, 1987), which involves a starting point, a sequence of locations connecting starting and ending points, and an ending point, is not universal. Mishra and Dasen (2008, p. 242) suggest, for example, that body-centred spatial notions of right and left and front and back are absent in some languages in which a geocentric frame of reference is preferred.

2.4. Previous studies on interaction in the tourism domain

To the author's knowledge, no other studies combine the phraseological approach to language with metadiscourse and the theories on spatial description. Some studies combine only two of the abovementioned methodological approaches; the most recent ones are described below.

Interactional metadiscourse is considered by Suau-Jiménez (2016, p. 206) as the result of three main variables, namely language, genre, and discipline. In her analysis of promotional English and Spanish tourism websites, she shows that both cultures achieve persuasion through a specific use of boosters (Suau-Jiménez, 2016, p. 216). Conversely, hedges are more frequent on English websites and are used by authors to filter their authority to make it sound more like a suggestion rather than an imposition. Readers are engaged by personal markers, particularly second-person pronouns, and self-mentions of the inclusive type, such as *we*, *us*, *our*, and *ours*. Other engagement markers are directives, which directly invite readers to carry out specific actions "based on a supposed level of trust and confidence established through other interpersonal markers and strategies" (Suau-Jiménez, 2016, p. 216). Interestingly, Suau-Jiménez's (2016) analysis of traveller forums reveals significant differences between English and Spanish from a metadiscursive perspective. For instance, hedges and personal pronouns are frequent in English, whereas attitude markers, self-mentions, and boosters are preferred by Spanish authors. As for stance and engagement, a stronger author's voice and authority are visible in Spanish travel forums, while on comparable English websites readers are constantly involved, giving the impression of peer-to-peer communication. These results confirm the strong interrelation existing between interpersonal markers and language, genre, and discipline.

Incelli's (2017) study on the promotional discourse of British, American, and Italian travel agencies combines a quantitative analysis of language with metadiscourse and explores the relationship between discursive patterns, interactional strategies, and culture. The results of this case study reveal that the three cultures adopt different promotional linguistic strategies. Thus, on British websites, readers' engagement is more highly represented, whereas in Italian the writer's identity is predominant. On the other hand, persuasion is achieved through direct address on British and American websites and through hedging in Italian.

Huang et al. (2020) use the interactional metadiscourse framework to conduct a quantitative and qualitative analysis of a corpus of English travel blogs. Results show that interactional devices are frequently used by bloggers in the construction of mutual interaction with the readers. Reader pronouns and boosters are the most used, followed by self-mentions and attitude markers. In general, adding up the percentages of all the interactional devices English travel blogs appear writer-oriented, and English bloggers seem inclined to express their stance and attitudes towards authentic personal travel experiences.

Unlike the case studies described above, this chapter aims to identify three different but strictly related types of interaction in British and Italian walking holidays' websites. More specifically, the focus is on the strategies to linguistically encode interaction between (1) words, patterns, and functions, (2) authors and readers in promotional tourist websites, and (3) participants and space in the route description.

3. Data and Methodology

The current study is based on a comparable corpus assembled by the author for the purpose of the present research. It includes websites advertising walking holidays in Italy and in the United Kingdom. The two subcorpora contain original Italian and English texts. They are comparable because they include texts dealing with the same topic, belonging to the same genre, and having the same communicative function. Texts were downloaded in 2021 by Google searching "vacanze a piedi" (literally, 'holidays on foot') for the Italian subcorpus, and "walking holidays in the UK" for the British subcorpus. The Italian Walking Holidays corpus (henceforth, ITWalHol corpus) contains 173,147 running words, and the UK Walking Holidays corpus (henceforth, UKWalHol corpus) contains 185,529 running words.

The websites included in the subcorpora advertise guided and self-guided walking holidays in Italy and in the UK. They usually provide a short description of the county or region where the walking tour starts and ends, followed by a detailed itinerary including the activities scheduled for each day. The description of the itinerary is usually followed by practical information such as accommodation options, useful and necessary equipment, the ideal time of year to join or try the tour, what is included and not included in the holiday price, and insurance options. For the two subcorpora, 25 different websites were selected, and, from each website, a series of itineraries ranging from five to 20 was downloaded and included in the corpus. The average length of the texts included in the corpus ranges from 500 to 700 words.

As aforementioned, the methodology adopted combines the analytical methods of Corpus Linguistics with the theories of metadiscourse and spatial description (Hyland, 2005; Taylor & Tversky, 1996). The analysis started by considering the wordlist generated by Sketch Engine[2] to identify the most frequent pronouns in the UKWalHol corpus. The personal pronoun *you* was the first in the list, and the personal pronoun *we* ranked second. These pronouns were selected as node words, because of their frequency and because they are the primary devices of the categories of engagement and self-mention. Furthermore, in the route perspective, speakers address listeners as "you", in that they become the changing viewpoint of spatial description. The two English pronouns' collocational profiles were analysed using Sketch Engine, and the discursive patterns and pragmatic associations that emerged were interpreted according to Hyland's (2005) and Taylor and Tversky's (1996) theoretical frameworks.

To compare and contrast the strategies identified in English, the analysis of the Italian subcorpus started from the most frequent combinations of the Italian translation equivalents of *you*, *we*, and *us*, namely *tu*, *ti*, *noi*, *ci*, *voi*, and *vi*. However, since in the Italian language subjects are not always expressed, all the verbs in the second-person singular and plural and in the first-person plural were analysed. Italian discursive patterns were also interpreted according to Hyland's (2005) and Taylor and Tversky's (1996) models.

In the last section of the analysis, the collocational profiles of the most frequent conjugated Italian verbs appearing in the wordlist were investigated. This further analysis aimed to identify patterns acting as temporal links that could be

---

2 http://www.sketchengine.eu [Last accessed: 28/09/2023].

considered as functionally similar to those identified in the collocational profiles of *you* and *we*. Similarities and differences between the two languages/cultures were described and discussed.

4. Analysis

As already described, the analysis aimed to check how the interaction between authors and readers is linguistically encoded on walking holidays' websites and how space is related to participants in the tour description. The analysis of the UKWalHol corpus focused on the pronoun *you* as a subject and object, on the subject pronoun *we*, and the object pronoun *us*. The analysis of the ITWalHol corpus started from the Italian translation equivalents of the English pronouns, namely, *tu*, *ti*, *voi*, *vi*, *noi*, and *ci*, and also focused on all the verbs in the second-person singular and plural and in the first-person plural occurring in the corpus.

4.1. The personal pronoun *you* in the UKWalHol corpus

The personal pronoun *you* occurs both as a subject and as an object pronoun 2,447 times (1.31 %). By analysing all its entries, it emerges that it is co-selected with verbs referring to the activities related to the holiday advertised and to the itinerary proposed. The frequent presence of *you* seems to suggest that the focus of description is on individual readers that are engaged through this form of direct addressing. An analysis of the patterns in which this pronoun is included may provide further insights and help better outline the focus of promotion.

Verbs in the active voice collocating with *you* as a subject mainly refer not only to the activity of walking (*walk, follow, pass*, etc.) but also to leisure activities (*enjoy, have time, discover, explore*, etc.).

The first two verbs with the highest frequency of association with *you* as a subject pronoun are modals, namely *can* (172 occurrences) and *may* (96 occurrences). Other modals in the list are *could, should, might*, and *must*. In most of their uses, modals are hedging markers, allowing information to be expressed indirectly as an opinion rather than a fact (Incelli, 2017, p. 75). According to Maci (2007, p. 58), *can* describes the idea of possibility, but, by giving the impression that the tourist can choose what to do, this modal, in some cases, may also express an off-record invitation, thus becoming deontic. Both *can* and *could* are followed by verbs describing further activities that can be carried out in the area where the tourist is or has arrived or by verbs emphasising a sense of achievement after a long day of walking. Interesting recurring patterns with *can* are *(from) here you can* (16 occurrences) and *(from) where you can* (51 occurrences). The pattern including *where* can also be observed when *you* is co-selected with *may* (nine occurrences) and with *could* (six occurrences), and it performs the function of giving suggestions. Examples (1), (2), and (3) represent these uses:

(1) Today you head to L'Eree by bus <u>where you can admire</u> Lihou island and even spend some time on the island, if you wish, <u>where you can explore</u> the ruins of a twelfth-century monastery.
(2) <u>Here, you can feel</u> a million miles from civilisation.
(3) Pass through the charming villages of Broad Campden and Blockley and descend to Batsford, <u>where you may visit</u> the arboretum and falconry centre.

The modal *may* is also used to describe the possibility that something is needed or to explain what tourists are allowed to do. This modal seems to act as a device of the category of hedges in that it indicates the subjectivity of the organisers' position and the degree of its negotiation, as exemplified in (4) and (5):

(4) <u>You may</u> require a taxi for the short distance from the guesthouse to the rail station.
(5) <u>You may</u> arrive at the guesthouse any time after 3 pm.

The modal *might* shows some interesting patterns of usage, as in 10 instances out of 18 it is used in the expressions *you might (also) be interested in* and *you might (also) like to*, describing further or alternative activities that can be enjoyed in the area or along the route. This is also a linguistic device belonging to the category of hedges in that it performs the function of negotiating a position.

Moving on to the modal *should*, it is used to give suggestions and opinions, or to describe a recommendation. In five instances it is co-selected with the expression *have time to*, to which we could add two instances including *arrive to find time to*. Some examples including the modal *should* are reported in (6) and (7):

(6) Just like the rivers along this trail, <u>you should explore and wander</u>, generally relax into the Dales Way!
(7) As the land is flat and this is a river walk, <u>you should consider</u> potential flooding of the path through the winter months.

Regarding the modal *must*, it collocates with *you* in five occurrences and is always included in the pattern *you must be (very) fit*, a recommendation that refers to those trails characterised by high difficulty levels. This modal is an example of modal obligation with the function of indirect directive (Quirk et al., 1985) and belongs to the category of engagement markers.

As noticed in the co-selections of *you* with *can*, *could*, and *may*, the adverbs *here* and *where* are included in patterns whose main function is indicating and suggesting possible activities and alternative options along the route. A closer look at the two adverbs associated with *you* reveals that the pattern including *here* occurs 53 times, whereas the one including *where* has 155 occurrences. More precisely, there are three possible patterns for each adverb co-selecting with *you*: apart from the modals *can*, *could*, or *may*, they can also include the auxiliary *will* or only a verb. Thus, the following patterns are reported:

a. *(from) here* + *you* + *can* + verb (17 occurrences)
b. *(from) here* + *you* + *will* + verb (23 occurrences)
c. *(from) here* + *you* + verb (13 occurrences)
d. *(from) where* + *you* + *can/may/could* + verb (61 occurrences)
e. *(from) where* + *you* + *will* + verb (37 occurrences)
f. *(from) where* + *you* + verb (57 occurrences)

When modals are included in the patterns, the aim of their usage is to give suggestions for the activities and attractions that can be enjoyed along the route or once the scheduled destination is reached, and mainly perform the instructional function. When *will* or only a verb is present, these patterns mainly perform the descriptive function and are used to explain how the itinerary is organised and to describe the different sections it includes. While modals are used as hedges to describe opinions on possible activities or attractions to visit, the auxiliary *will* acts more as a device of the category of boosters emphasising certainty.

Moreover, the adverbs *here* and *where* are used as linguistic devices to give cohesion to the description of routes by following the given/new principle. Indeed, the two adverbs refer to known spatial locations, while new landmarks are mentioned in the remaining part of the sentence, thus establishing a coherent and cohesive relation among them.

On the other hand, in 168 instances, *you* occurs in a pattern including *as* followed by verbs of motion, such as *approach, climb, cross, descend, follow, head, pass*, and *walk*, to mention just the most frequent ones. Examples are (8) and (9):

(8) You will largely follow the Erme Estuary and enjoy views over to Burgh Island <u>as you approach</u> Bigbury-on-Sea.
(9) <u>As you walk,</u> you'll pass by a number of old mills, lades, and weirs.

This pattern is used to describe what you will see, enjoy, or experience while following the path, thus giving the idea of an immersive walking experience. This confirms the tendency towards a description based on the route perspective, which implies the presence of two events linked in time (enjoy-approach; walk-pass), in this case by means of the temporal link *as*.

Another frequent pattern occurring 30 times includes *before* followed by *you* and by verbs of motion, among which the most frequent are *arrive, continue, leave*, and *reach*. This pattern is usually employed when suggestions for further activities to be carried out along the route are given or when the activities to do or the attractions visible before a specific section of the path are described, as shown in example (10):

(10) Enjoy some of the town's history and heritage before you leave.

Interestingly, the adverb *after* is never used with the same function as *before*, while a similar behaviour can be observed with *once* but in only four instances. In this case, the temporal link *before*, which connects two events as typical of route descriptions, always introduces the given information although occurring, most of the times, in the second part of the sentence. This change of position of given and new elements may be considered a strategy to emphasise the new element, which, in example (10), is the attractions and sights not to be missed.

As already noticed when the adverbs *here* and *where* have been discussed, the auxiliary *will* is also frequently co-selected with *you*. In 410 instances, *will* is used to describe future activities, thus performing the instructional function, or acts as a booster expressing certainty and promise that something will be found or seen, mainly performing the persuasive function by stimulating desire (cf. Manca, 2016a).

In 70 occurrences, the subject pronoun *you* occurs with verbs in the passive form, among which the most frequent ones are *collect* (17 occurrences), *reward* (15 occurrences), *transport* (10 occurrences), and *take* (eight occurrences). Apart from *reward*, all of them refer to technical details related to the holiday, to when and where participants will be collected, or how they will be transported or taken to their accommodation or the station. Conversely, *rewarded* has an interesting collocational profile and is used to describe what walkers will find after a challenging section of the route or at their destination. In almost all cases, *reward* is associated with *views* occurring in a prepositional phrase introduced by *with*, as in *Today you are rewarded with amazing views from the offset*. Verbs in the passive form seem to suggest that the writer's presence in the text tends to be avoided in favour of a focus on readers. Self-mention options are, therefore, replaced by engagement markers.

When *you* is used as an object pronoun, it occurs with verbs referring to the technical aspects of the holiday or of the itinerary. Among the most frequent verbs, there is *take*, which is mainly used in the patterns *take you to*, *take you through*, *take you past*, *take you into*, *take you along*, and *take you across*. As predictable, subjects of *take* are all items referring to the route and are *route*, *path*, *trail*, *walk/walking*, *way*, *climb*, and *ascend/descend* mainly when the verb is in the third person singular, *taxi* when the verb is preceded by *will*, and *we* when the verb is in the Present Simple. The presence of nouns referring to the path in the subject position does not imply a change of perspective in spatial description, because the presence of *you* suggests that the viewpoint is still the listener. The patterns including *route, path, trail, walk/walking, way, climb*, and *ascend/descend* are also examples of fictive motion (Talmy, 2000), that is to say, of cases in which an entity is depicted as moving even though it is static in the real world (Cappelli, 2012, p. 8).

Summarising the results obtained from the analysis of *you*, three groups of verbs can be identified in association with this pronoun as a subject: (1) verbs referring to the activity of walking, (2) verbs referring to leisure activities, and (3) modals (*can, may, could, might, must*). Modals are mainly used to give suggestions and to negotiate the author's position and, when they perform this function, belong to the category of hedging markers. Other interesting associations with the subject *you* are those including the adverbs *here* and *where*. The pattern *(from) here/where* + *you* + *can/may/could* + verb is used to describe and give suggestions about the activities, sights, and attractions that can be enjoyed at a specific section of the tour, and, for this reason, it performs the function of a hedging marker. Conversely, the pattern *(from) where* + *you* (+ *will*) + verb describes the certainty of finding something and acts as a booster. The adverbs *here* and *where* associated with *you* are also frequently used as devices of spatial descriptions. They refer to known locations along the itinerary and thus contribute to a cohesive alternation of given and new elements in the description. Other devices of spatial description connected with *you* as a subject are the patterns *as you* + verb of motion, which is used to describe events linked in time, and *before you* + verb, which is not only used to provide suggestions for activities and attractions to walkers but also to introduce the given element in the sentence.

As for verbs in the passive form occurring with *you*, they mainly refer to technical details related to the holiday and reduce the writer's presence shifting the focus on readers. When *you* is used as an object, it is associated with verbs describing the organisation of the walking holiday.

The analysis described in this section started with the identification of the discursive patterns including the personal pronoun *you*, an item that was selected based on its high frequency of occurrence in the UKWalHol corpus and on its relevance from metadiscursive and spatial perspectives. However, to explore further the interactional and spatial perspectives adopted in the promotion of British and Italian walking holidays, the analysis continues with the identification of the discursive patterns of another frequently used personal pronoun, the first-person plural pronoun *we* and its corresponding object pronoun *us*.

4.2. The personal pronouns *we* and *us* in the UKWalHol corpus

The pronoun *we* occurs 745 times in the UKWalHol corpus, which corresponds to a percentage of 0.4 % with respect to the total number of words in the corpus. It is used both as an excluding and including *we* but with different collocational ranges. When it is used as an excluding *we*, the pronoun refers to the holiday organisers and is a device of self-mention. It is co-selected with verbs referring to the organisation of the vacation and to the role performed by the organisers as advisers willing to take care of the walkers' satisfaction. Verbs collocating with the excluding *we* are *advise, arrange, book, collect, organise, guide, have (lots of options), offer, provide, transfer, hope (you enjoy/have enjoyed), look forward, suggest*, and *delight, love, like* but only in the conditional form *would be delighted/would love/would like to*. Some examples are reported in (11), (12), and (13):

(11) You will make use of the notes <u>we will provide</u> you with and <u>we arrange</u> your accommodation on a B&B basis.
(12) If the cloud is low, <u>we recommend</u> taking the low-level route.
(13) <u>We hope</u> you enjoyed your stay and <u>we look forward</u> to welcoming you again soon.

The verb *recommend* is an example of indirect directive (Quirk et al., 1985) and belongs to the category of engagement markers.

However, in most cases, the pronoun is used as an including *we* and refers to the whole group of people who will join the walking holiday, hence acting as an engagement marker. The verbs which are frequently co-selected with it refer to the activities of walking and exploring, such as *climb, continue, cross, descend, drive, enjoy, explore, follow, head, pass, return, see, start, visit*, and *walk*.

The pronoun *we* is also co-selected with the modal *can* in 36 instances and with the modal *may* in 10 instances, while it never occurs with *might*. Interestingly, when it occurs with *can* as an including *we* (10 times), it is often preceded by the adverbs *here* and *where*, which have already been observed in association with *you*. These adverbs are co-selected with *we*, also when *can* is not present and the adverb *there* replaces *here* in some instances. The pattern *(from) here/there we (can)* + verb occurs 14 times, whereas *(from) where we* + verb occurs 31 times. The presence of these patterns in the collocational profile of this pronoun suggests that they are a structural feature of route descriptions aiming at establishing cohesion. They also perform an instructional function as they are used to give further suggestions for the activities and sights available in the area and are used as hedging markers.

In 25 instances, the pronoun *we* is included in a pattern that has already been noticed with *you*, that is to say, *as we* + verb, whose function is to describe what will be enjoyed along the trail to convey the idea of an immersive walking experience. The patterns *as* + known activity + new activity/attraction or new activity/attraction + *as* + known activity are used to make the text more cohesive and to emphasise new information. Examples (14) and (15) show this:

(14) <u>As we approach</u> the Malvern Hills from the north, we pass through the village of Alfrick.
(15) Some sightseeing in the morning, <u>as we walk</u> up the Sabbath Walk route to Coalbrookdale.

In conclusion, the first-person plural pronoun is used both as an excluding and as an including *we*. When used with an exclusive function, it belongs to the category of self-mention and it is associated with verbs semantically referring to the organisation of the holiday. When it performs the inclusive function, it performs the role of an engagement marker and occurs with two groups of verbs: (1) verbs referring to the activities of walking and exploring, and (2) modals. The pronoun *we* also occurs in the pattern *(from) here/there/where we (can)* + verb, used to describe suggestions for further activities and sights and to create cohesion in the description, and in the pattern *as we* + verb, used to link events and to give the idea of an immersive walking holiday.

To balance this analysis with that of the personal pronoun *you*, the object pronoun *us* was also investigated, although its frequency in the corpus is not high (75 occurrences corresponding to 0.04 %). In its excluding function, this object pronoun is used as a self-mention marker and occurs with verbs in the Imperative form in expressions such as *please ask us, (please) contact us*, and *just/please let us know*. Therefore, self-mention and engagement markers can be observed in the same expression. When *us* is used as an including pronoun, it frequently occurs with the verbs *take* and *bring*. As observed with the object pronoun *you*, the subjects of the expressions including *us* and the verbs *take* and *bring* are *route, path, drive, trail*, and *walk*, and the type of motion described is fictive (Talmy, 2000).

The analyses of *you* and *we* in the UKWalHol corpus revealed a clear tendency towards a promotion focused on readers/potential customers. Hedges are preferred to boosters, and engagement markers are more frequently used than self-mention markers. From a spatial perspective, cohesion is created by alternating given and new elements, which can be positioned either in the first or in the last part of the sentence. Furthermore, the use of expressions including the conjunctions *as* and *before* help create temporal links between actions or give the idea of an immersive walking experience with plenty to do and see.

To check if a change in language and culture within the same discipline (Suau-Jiménez, 2016) involves different perspectives not only in the use of interactional devices but also in the features of spatial description, in Section 4.3. the analysis aims to identify if and how the above-described linguistic options are present in the ITWalHol corpus.

4.3. Pronouns and verbs in the ITWalHol corpus

Personal pronouns are not always expressed in Italian because verb suffixes allow language users to identify subjects. A search in the ITWalHol corpus for the personal pronouns *tu*, *voi*, and *noi*, which are the translation equivalents of the English *you* and *we*, reveals, as expected, that these items display a very low frequency. For this reason, the analysis focuses on all those verbs in the corpus that are conjugated in the second-person singular and plural and in the first-person plural. Table 1 summarises the total number of occurrences and the percentages of occurrence of verbs conjugated in the second-person singular, in the second-person plural, and in the first-person plural. As already done in the previous analyses, percentages are calculated with respect to the total number of words occurring in the subcorpus under analysis in order to make comparisons across the two subcorpora easier and more accurate.

Table 1. Occurrences and percentages of verb forms in the ITWalHol corpus

	second-person singular	second-person plural	first-person plural
Number of occurrences	44	543	1,648
Percentage of occurrence	0.02 %	0.31 %	0.95 %

The percentages reported in Table 1 suggest that verbs in the first-person plural have a frequency of occurrence three times higher than the frequency of verbs in the second-person plural. Although most of the verbs in the first-person plural refer to an including *we* and, therefore, to readers/participants, the Italian promotion seems to address readers more as members of a group rather than as individuals. Spatial description adopts the route perspective as suggested by the presence of participants moving along a path to reach a goal. The verb tenses with the highest frequency of usage are the Present Simple Indicative, the Future Simple Indicative, and less frequently the Imperative. It may be hypothesised that Italian texts are more descriptive than directive, although an analysis of verbs is needed to confirm this hypothesis.

4.3.1. Second-person pronouns and verb forms

Verbs occurring in the second-person singular are *vedere* ('see'), *scoprire* ('discover'), *potere* ('can'), and *vivere* ('live'), whereas verbs occurring in the second-person plural are 23 and refer to activities related to the walking holiday, such as *raggiungere* ('reach'), *arrivare* ('arrive'), *scendere* ('descend'), *entrare* ('enter'), and *lasciare* ('leave'), to mention just the most recurrent ones.

The most frequent verb in the second-person plural is *potere* ('can'), mainly used in the Present Simple Indicative *potete* and in the Future Simple Indicative *potrete*. Frequent verbs co-selected with these two forms of the modal are *ammirare* ('admire'), *fare (un giro/una pausa)* ('take a tour/a break'), *scegliere* ('choose'), *utilizzare* ('use'), and *visitare* ('visit'). All these verbs refer to activities and attractions that can be enjoyed during the walking holiday. As observed with the English modal *can*, *potere* is used as a device of the category of hedges aimed at describing possibilities rather than certainties. The adverbs *dove* ('where') and *qui* ('here'), which were frequently included in patterns with *you* and other modals in the UKWalHol corpus, are also present here (cf. example (16)), although with a lower frequency:

(16) <u>Qui potete</u> visitare le bellezze naturali e storiche di Taormina ('Here you can visit the nature and heritage attractions of Taormina').[3]

A look at the concordances of the adverbs *dove* and *qui* reveals that these items occur, respectively, 35 times and 28 times with verbs in the second-person singular. They show a more frequent association with verbs in the first-person plural (cf. Section 4.3.2.) and in the impersonal forms of verbs (e.g., *Da qui <u>si può</u> ammirare la grotta Su Forru* – 'From here one can admire Su Forru cave'). Their usage is linked to the given/new principle in that they are used to refer to the known landmark, thus helping create cohesion in the text.

The other modal in the second-person plural is *volere* ('want') in the Present Simple Indicative *volete*. It occurs only six times but, in almost all cases, it occurs in an expression including the conjunction *se* ('if'), hence performing a function similar to the English expression *if you wish*, in which what is proposed is signalled, through the category of hedges, as open to negotiation.

There are no other recurring patterns in the collocational profiles of verbs in the second-person plural, and expressions functionally similar to the English

---

3 The back translations provided are as literal as possible to allow readers to understand the lexical, syntactic, and stylistic choices in the Italian examples.

as you + verb and *before you* + verb were not identified. Therefore, it may be hypothesised that those functions might be performed differently in Italian.

The object pronouns *ti* and *vi* ('you') have also been analysed. The second-person singular *ti* occurs 53 times (0.03 %) and frequently collocates with the verb *aspettare* ('wait'). The plural *vi* has 200 occurrences (0.11 %) and collocates with the verbs *portare* ('take'), *condurre* ('lead'), *fare* ('make'), and *lasciare* ('leave'). Object pronouns confirm the same tendency already noticed with subject pronouns, that is, the preference to avoid second-person singular pronouns to address the audience. However, before commenting on the strategies used by Italian websites to engage readers, the analysis continues with the identification of the discursive patterns including first-person plural pronouns and verbs conjugated in the first-person plural.

4.3.2. First-person plural pronouns and verb forms

In the ITWalHol corpus, 58 verbs are conjugated in the first-person plural, mainly in the Present Simple Indicative and in the Future Simple Indicative. The range of verbs conjugated in the first-person plural is varied and includes verbs referring to the route (*raggiungere*/'reach', *proseguire*/'continue', etc.), to the type of walking (*salire*/'climb', *camminare*/'walk', etc.), and to technical details of the holiday (*dormire*/'sleep', *consigliare*/'recommend', etc.). The modal *potere* ('can') is not the most frequent verb in the list and this is interesting, particularly if we consider that this modal had the highest number of occurrences both in the lists of the Italian verbs in the second-person singular and plural and in the lists of verbs co-selected with the English *you* and *we*. The Italian modal *potere* ('can') is used in the Future Simple *potremo*, in the Present Simple Indicative *possiamo*, and less frequently in the Present Conditional *potremmo*. The verbs more frequently co-selected with this modal are *ammirare* ('admire'), *salire* ('ascend'/'climb'), and *mangiare* ('eat'), the last one less frequently than the first two. The adverbs *dove* ('where') and *qui* ('here') are present when the verb is in the Future Simple: *dove* occurs 12 times and *qui* six times out of 45 occurrences. The function of the expressions including these adverbs is the same as that identified in the previous analyses, both in English and Italian, that is to say, suggesting further attractions and activities to be explored and enjoyed and creating cohesion by linking new and given information. Furthermore, the future form of the verbs suggests that these expressions are strategically used as boosters.

The analysis of the collocational profile of all the verbs in the first-person plural does not reveal any further interesting patterns of usage. As for the object pronoun *ci* ('us'), occurring 293 times, its percentage of occurrence (0.17 %) confirms a clear preference for usage related to the category of engagement markers. The most frequent verbs co-selected with *ci* are *accompagnare* ('accompany'), *aspettare/attendere* ('wait'), *condurre* ('lead'), *fare* ('make'), *permettere* ('allow'), and *portare* ('take'). This pronoun is always used with an including meaning and describes the tendency of Italian websites to address readers as members of a group rather than as single individuals. This way of addressing mitigates the directness observed in the UKWalHol corpus and, although suggesting a reader-centred perspective, it is strategically used to establish a distance with readers.

It must be pointed out that, in the Italian corpus, the impersonal form expressed by *si* combined with the third person singular of verbs is frequently used to address readers. This form is another indirect form of addressing and contributes to placing emphasis on the activity expressed by verbs rather than on the subjects. According to Incelli (2017, p. 83), this generic impersonal structure creates a vague 'you' and describes something which is programmed rather than a novel experience. The limited use of directives that would be too direct for an Italian audience (cf. Manca, 2016a) confirms the tendency of Italian websites to establish a distance between authors and participants.

4.3.3. Other frequent verbs

The results obtained from the Italian subcorpus revealed the absence of expressions that were functionally similar to the English *as you* + verb and *before you* + verb. Therefore, it may be hypothesised that, in Italian, those functions might be performed differently, for example, by verbs rather than by conjunctions, as we will see below. For this reason, in this section, we focus on the most frequent verbs appearing in the wordlist of the ITWalHol corpus.

Interestingly, among the first 12 most frequent conjugated verbs, there are eight non-finite verbs, namely five infinitives and three gerunds. They are *visitare* ('visit' – 124 occurrences), *raggiungere* ('reach' – 123 occurrences), *arrivare* ('arrive' – 101 occurrences), *passando* ('passing' – 101 occurrences), *ammirare* ('admire' – 84 occurrences), *scoprire* ('discover' – 76 occurrences), *camminando* ('walking' – 69 occurrences), and *seguendo* ('following' – 69 occurrences). A look at the collocational profiles of these verbs may help us gain further insights into Italian discursive patterns and their specific functions.

The verb *visitare* ('visit') frequently occurs with the word *tempo* ('time') in expressions such as *avere tempo (libero) per visitare* + attraction ('have (free) time to visit + attraction'), *è possibile/possibilità di visitare* + attraction ('it's possible/possibility to visit + attraction'), *non dimenticate di visitare* + attraction ('don't forget to visit + attraction'), and *vale la pena visitare* + attraction ('worth visiting

+ attraction'). All these expressions aim to provide suggestions for attractions worth visiting or to invite walkers to visit the most interesting highlights of the area. This confirms that providing suggestions for further activities or attractions to be enjoyed in the area is a recurring feature also in the Italian websites. It may be assumed that the instructional function is a feature of both the British and the Italian discourses of walking holidays, although differently encoded in the two languages. It is interesting to notice that suggestions for attractions to visit are given through hedges or engagement markers, such as direct and indirect directives, conveying different degrees of interaction.

With regard to the verb *raggiungere* ('reach'), it occurs 123 times and in 50 cases it is preceded by the preposition *per* ('to'), which explains the reason and the aim of the action described by the preceding verb or which describes what comes next in the route, as in example (17):

> (17) *Iniziamo a salire per raggiungere la sommità del Monte Rosso* ('We start to climb to reach the peak of Mount Rosso').

In 34 cases, this verb is included in the expression *fino a raggiungere* ('until + reach'), preceded by motion verbs, as in *Da qui riprenderete il Sentiero degli Dei fino a raggiungere Nocelle* ('From here, you will take again the Sentiero degli Dei trail until [you] reach Nocelle'). A less frequent expression is *prima di raggiungere* ('before [you] reach'), occurring six times.

Regarding the verb *arrivare* ('arrive'), in the infinitive it has the same collocational profile as *raggiungere* ('reach'). It occurs in the expressions *per arrivare, fino ad arrivare*, and *prima di arrivare*. The patterns *per/fino a/prima di* + infinitive ('to/until/before + infinitive') perform a cohesive role: they introduce the known spatial location and contribute to emphasising the new landmarks described in relation to them.

As already observed, the infinitive *ammirare* ('admire') is strongly linked to the concept of 'possibility' as visible in its frequent association (62 occurrences out of 84) with the modal *potere* ('can'), and with the expression *è possibile ammirare* ('it is possible to admire'). These patterns are examples of hedges where the activities proposed are presented as open to negotiation. What is interesting here is the presence of a stative verb at the top of the frequency list. This confirms that Italian websites of walking holidays tend to provide descriptions rather than directions to readers.

The last infinitive in the group of the most frequent verbs in the ITWalHol corpus is *scoprire* ('discover'). As the other infinitives already analysed, it also shows a frequent association with the preposition *per* ('to'), describing the aim of the action expressed by the preceding verb or explaining what comes after a stop, a walk, or a transfer. It is frequently used in the expression *da scoprire* ('to be discovered'), always referring to an attraction to be discovered and not to be missed. This can be considered an example of an indirect directive implying the idea of a recommendation. The verb *scoprire* ('discover') also shows a frequent association with the verbs *portare* ('lead') and *fare* ('make'), both describing something which will allow walkers to discover stunning attractions and places.

The last three Italian verbs worth analysing due to their high frequency are the gerunds *passando* ('passing'), *camminando* ('walking'), and *seguendo* ('following'). First, the verb *passando* ('passing') occurs mainly with the prepositions *da* ('from') and *per* ('through') in expressions that suggest a point of arrival passing through a place. However, the place already known to walkers is not always mentioned before the new one and the given/new sequence is not as systematic as it was with *before* in the English corpus. In example (18), the ridge (*crinale*) is the element that is already known while *Poggio della Pagana* is the new element. On the contrary, in example (19), the final destination *Sonnenhof* is mentioned after the two villages which represent the new information. In conclusion, the verb *passando* ('passing') always introduces a new element, but it does not necessarily occupy the last part of the sentence. It is also used as a device that helps create a temporal link in the narration.

(18) *Si continua l'escursione lungo il crinale <u>passando da Poggio della Pagana</u> che regala una meravigliosa vista* ('One continues the excursion along the ridge passing from Poggio della Pagana which offers a wonderful view').

(19) <u>*Passando sopra i paesini di Castelbello e Ciardes*</u> *saliremo piuttosto rapidamente fino al Sonnenhof* ('Passing above the villages of Castelbello and Ciardes, we will rapidly ascend until Sonnenhof').

Furthermore, the frequent presence of impersonal forms of verbs (example (18)) shifts the focus of the description on the route rather than on the walker.

Secondly, a similar function is performed by *camminando* ('walking') which, due to its frequent co-selection with impersonal forms of verbs, focuses on the route and on how and through which places the different sections of the route can be reached. It is almost always associated with prepositions, such as *lungo* ('along'), *tra* ('among'), *su/sul* ('on/on the'), and *in/nel* ('in/in the').

Last but not least, the verb *seguendo* ('following') is usually associated with items referring to the route to be followed to reach a destination (*percorso*/'route', *sentiero*/'path'). In many cases, the route is described in positive terms, employing qualifying adjectives or nouns, referring to the beautiful and varied landscape walkers will go through or focusing on the history behind the trail.

The above-described gerunds perform the role of temporal linkers in that they describe two actions happening simultaneously and help create cohesion by introducing the new elements of the message. For this reason, they could be considered pragmatic equivalents of the English *as* and *before*, as illustrated in examples (20), (21), and (22):

>(20) <u>Si rientra</u> nel Parco Regionale Naturale di Porto Selvaggio e Palude del Capitano, <u>passando</u> dalle grotte preistoriche di Uluzzo e dalla splendida baia di Porto Selvaggio ('One comes back to the nature reserve of Porto Selvaggio and Palude del Capitano, passing from the Uluzzo prehistoric caves and from the stunning bay of Porto Selvaggio').
>(21) <u>Scendiamo</u> sin verso il Vallone di Vallanta <u>camminando</u> nel bosco ('We descend towards the valley of Vallanta walking in the woods').
>(22) <u>Usciremo</u> dal paese di Colle <u>seguendo</u> il meraviglioso Sentierelsa ('We will go out of the village of Colle following the beautiful Sentierelsa').

5. Conclusions

The present analysis has described the discursive patterns including personal pronouns and verbs in a comparable corpus of British and Italian websites advertising walking holidays. The analysis aimed to identify frequent linguistic strategies systematically used to create interaction between authors and readers and to describe how space and participants are related to each other in this web genre. The results illustrate some of the promotional features of British and Italian walking holidays from three perspectives: phraseological, metadiscursive, and spatial.

Starting from the similarities identified across the two subcorpora, both the British and Italian websites adopt the route perspective to describe walks, as the frequent occurrence of personal pronouns suggests. This implies the use of devices which contribute to the creation of cohesion, as observed with the English adverbs *here* and *where*, and the Italian *dove* ('where') and *qui* ('here'). All these adverbs are related to the known element and introduce the sentence including the new element of the message.

In both corpora, modals such as *can, could, may, might*, and *should* in English and *potere* ('can') in Italian mainly perform the function of hedges and are used by authors/organisers to suggest activities to do and enjoy, or attractions to visit and explore. Activities and attractions are presented as possibilities open to negotiation.

What the English and the Italian websites have in common but encode differently are temporal links. A route perspective describes at least two events linked in time, and this is visible in the way *as* and *before* are used in the English corpus. In the Italian corpus, activities happening at the same time are not described by using conjunctions but by using verbs in the gerund form. Both the English conjunctions and the Italian gerunds are also devices to create cohesion by allowing the alternation of given and new elements.

However, the two groups of websites differ in the way interaction between authors/holiday organisers and readers/participants is established. In English, the personal pronoun *you* displays a higher frequency than *we* (1.31 % vs 0.4 %) and its collocational profile describes the centrality of walkers in this type of holiday. This is further confirmed by the presence of about 1,500 verbs in the second-person singular and plural of the Imperative, which in Hyland's (2005) model are considered engagement markers. In Italian, the percentage of occurrence of second-person singular and plural pronouns and verb forms is 0.33 %, while first-person plural pronouns and verb forms occur with a percentage of 0.95 %. These figures explain that, in Italian, the focus of promotion is the walker but as a member of a group. Although participants are seen as the focus of promotion in both subcorpora, in Italian the frequent use of first-person plural pronouns and verb forms mitigates the directness conveyed by second-person singular and plural pronouns and verb forms. Furthermore, the discursive patterns identified in the ITWalHol corpus as well as the frequency of impersonal forms of verbs and the limited presence of imperatives suggest that the distance between authors and readers is greater than in English. The Italian descriptions seem to be more characterised by an exchange based on "giving" rather than on "demanding" (Eggins, 2011, p. 144; Manca, 2016b, p. 14), the latter being more a feature of the English promotion.

Furthermore, from a spatial perspective, the high frequency of stative verbs, such as *ammirare* ('admire'), and the limited use of imperatives give the Italian websites the rhetorical structure of a description of route. By contrast, in English the frequent presence of imperatives involving "physical acts" (Hyland, 2005) and of motion verbs may suggest the choice of a style that alternates between the description of routes and the description of directions.

Finally, it is interesting to notice that the three most frequent verbs in English, apart from *can*, are *take*, *follow*, and *pass*, which refer to walking. In Italian, *raggiungere* ('reach'), *arrivare* ('arrive'), and *visitare* ('visit') are at the top of the list. These occurrences seem to suggest that, in the Italian promotion, the goal is more important than the path and the perspective adopted is projected towards the destination.

The limits of this study are evident in that further analyses would be needed to define more precisely the features of the British and Italian promotions of walking holidays. Nevertheless, the differences and the similarities observed in the two styles of promotion may provide the basis upon which scholars and tourism experts can develop a model that takes into account linguistic, stylistic, and cultural differences for an effective promotion across cultures and languages. Furthermore, the discursive patterns identified and their pragmatic meanings may have pedagogical implications for students specialising in tourism language and translation who need to become aware of the strong interaction existing between language, disciplines, genres, and cultures.

References

Buhler, K. (1982). The deictic field of language and deictic words [Abridged translation of K. Bühler (1934)]. In R. J. Jarvella & W. Klein (Eds.), *Speech, place and action* (pp. 9–30). Wiley.

Cappelli, G. (2012). Travelling in space: Spatial representation in English and Italian tourism discourse. *Textus, 1*, 51–67. https://doi.org/10.7370/71234

Eggins, S. (2011). *An introduction to systemic functional linguistics* (second edition). Continuum.

Fulga, A. (2012). Language and the perception of space, motion and time. *Concordia Working Papers in Applied Linguistics, 3*, 26–37.

Huang, Y., Wang, H. J., & Tang, J. H. (2020). A study of interactional metadiscourse in English travel blogs. *Open Journal of Modern Linguistics, 10*, 785–793. https://doi.org/10.4236/ojml.2020.106048

Hyland, K. (2005). *Metadiscourse: Exploring interaction in writing*. Continuum.

Incelli, E. (2017). A cross-cultural contrastive analysis of interpersonal markers in promotional discourse in travel agency websites. In G. Palumbo (Ed.), *Testi, corpora, confronti interlinguistici: approcci qualitativi e quantitativi* (pp. 65–86). EUT Edizioni Università di Trieste. https://doi.org/10.13137/978-88-8303-913-3/18481.

Lakoff, G. (1987). A cognitive theory of metaphor. *Philosophical Review, 96*(4), 589–594.

Levelt, W. J. M. (1984). Some perceptual limitations on talking about space. In A. J. van Doorn, W. A. van der Grind, & J. J. Koenderink (Eds.), *Limits on perception* (pp. 323–358). VNU Science Press.

Levelt, W. J. M. (1989). *Speaking: From intention to articulation*. MIT Press.

Levinson, S. (1996). Frames of reference and Molyneux's question: Cross-linguistic evidence. In P. Bloom, M. A. Peterson, L. Nadel, & M. Garrett (Eds.), *Space and language* (pp. 109–169). MIT Press.

Maci, S. (2007). Virtual touring: The web-language of tourism. *Linguistica e Filologia, 25*, 41–65.

Manca, E. (2004). The language of tourism in English and Italian: Investigating the concept of nature between culture and usage. *ESP Across Cultures, 1*, 53–65.

Manca, E. (2016a). *Persuasion in tourism discourse: Methodologies and models*. Cambridge Scholars Publishers.

Manca, E. (2016b). Official tourist websites and the cultural communication grammar model: Analysing language, visuals, and cultural features. *Cultus, 9*(1), 2–22.

Mishra, R. C., & Dasen, P. R. (2008). Spatial language and concept development: Theoretical background and overview. In N. Srinivasan, A. K. Gupta, & J. Pandey (Eds.) *Advances in cognitive science* (Volume 1) (pp. 240–252). Sage Publications.

Quirk, R., Greenbaum, S., Leech, G., & Svartvik, J. (1985). *A comprehensive grammar of the English Language*. Longman.

Scott, D. (2021). Sustainable tourism and the grand challenge of climate change. *Sustainability, 13*(4), 1966. https://doi.org/10.3390/su13041966

Sinclair, J. (1991). *Corpus, concordance, collocation*. Oxford University Press.

Sinclair, J. (1996). The search for units of meaning. *Textus, IX*(1), 71–106.

Suau-Jiménez, F. (2016). What can the discursive construction of stance and engagement voices in traveler forums and tourism promotional websites bring to a cultural, cross-generic and disciplinary view of interpersonality? *Ibérica, 31*, 199–220.

Talmy, L. (2000). *Towards a cognitive semantics* (Volume 2). MIT Press.

Taylor, H. A., & Tversky, B. (1996). Perspective in spatial descriptions. *Journal of Memory and Language, 35*(3), 371–391. https://doi.org/10.1006/jmla.1996.0021

Tversky, B. (2004). Narratives of space, time, and life. *Mind & Language, 19*(4), 380–392. https://doi.org/10.1111/j.0268-1064.2004.00264.x

Acknowledgements

I would like to thank my colleague Gloria Cappelli for her helpful suggestions on an earlier draft.

Macarena Palma Gutiérrez

Universidad de Córdoba

l82pagum@uco.es

9 Syntactic alternations with verbs of motion: A corpus-driven analysis of the language of adventure tourism

Abstract: This chapter analyses the syntactic and semantic characterisation of motion verbs, paying particular attention to the structural and relational link between this type of verbs and their participants in terms of argument structure realisation, as well as the syntactic alternations displayed by these verbs in the specialised domain of adventure tourism. Particularly, two subclasses of verbs of motion are examined: the subclass of "run" verbs (e.g., *hike*) and the subclass of verbs that are vehicle names (e.g., *canoe*), as some of the members of these groups were attested in previous studies as the most productive predicates in this specialised discourse (cf. Durán-Muñoz & L'Homme, 2020). The methodology followed was corpus-driven combined with a syntax-semantic approach. The corpus consulted was ADVENCOR (Durán-Muñoz & Jiménez-Navarro, 2021), a specialised corpus focused on the field of adventure tourism. After a process of manual discarding, the sample of instances of grammatical constructions incorporating the selected verbs consisted in 578 contextualised instances. The sample was classified into different groups, depending on the subtype of verb of motion incorporated and the type of syntactic alternation displayed in each case. The main findings show that, although the intransitive basic or unmarked forms are frequent, some verbs tend to occur more productively with other syntactically derived alternations. This shows that the verbs under study exhibit a more restricted behaviour when used in the language of adventure tourism than their counterparts in general domains.

Keywords: adventure tourism, basic or unmarked form, derived or marked form, motion verb, syntactic alternation

1. Introduction

Within the field of terminology, it has traditionally been assumed that nouns are the parts of speech that prototypically convey more terminological value than the other units (Sager, 1990; Vargas Sierra, 2012). This idea is based on the implementation of approaches that consider nouns as the starting point of their terminological analysis. According to linguists like Levin (1993), Hale and

Keyser (1998), and Fellbaum (1990), verbs are the parts of speech that provide information about the arguments (or participants) involved in the action denoted, as they indicate the structural, relational, and semantic framework for the sentences in which they occur. Therefore, verbs provide information about their argument structure realisation, including the order and number of participants involved in the action denoted by the predication. Changes in the argument structure of a verb involve different syntactic alternations (Levin, 1993), which, in turn, imply differences in meaning at a lexico-semantic and discourse-pragmatic level. In addition, researchers like Casademont (2014), Buendía-Castro (2012, this volume), Jacinto García (this volume), L'Homme (1998, this volume), and Durán-Muñoz and Jiménez-Navarro (2023) consider verbs as conveyances of knowledge that help characterise specialised domains.

The present study also considers verbs as specialised units of language in specific fields, and, hence, it aims at contributing to preceding studies that share this idea. Along with the lines of Durán-Muñoz and Jiménez-Navarro (2023) and Durán-Muñoz and L'Homme (2020), this study focuses on the specialised domain of adventure tourism, which can be defined as the "type of tourism [that] involves the tourist's active participation to create a real adventure experience in nature" (Durán-Muñoz & Jiménez-Navarro, 2023, p. 28). In this specialised field, verbs (particularly, those describing actions carried out in adventure activities like *trek* and *raft*) play an important role in discourse, as further detailed in this work.

In Durán-Muñoz and L'Homme (2020, p. 43), motion verbs are considered the most frequently found verbs in the specialised language of adventure tourism. Motion verbs are those that describe the spatial displacement of an entity, being either animate or inanimate. In their study, Durán-Muñoz and Jiménez-Navarro (2023) also demonstrate that, among the class of motion verbs, real motion verbs are much more productive than fictive motion verbs in this specialised discourse. Real motion verbs involve a type of entity displacement that is real or literal; thus, there is an animate entity and its "real movement through physical space" (Matlock, 2004, p. 14), as in *You crossed the valley* (ADVENCOR; cf. Section 3). On the other hand, fictive motion verbs comprise an abstraction of motion in which the viewer's perception depicts a motionless inanimate entity as moving through physical space (Langacker, 1986), as in *The trail crossed the valley* (own elaboration).

Since real motion verbs are the most common type of verbs found in the specialised field of adventure tourism, the present corpus-driven study focuses on their examination and frequency of occurrence with regard to the multiple syntactic alternations in which they can be found. Thus, we analyse six verbs that were selected due to their productivity and relevance in the domain under study, as demonstrated in Durán-Muñoz and Jiménez-Navarro (2023). Following Levin's (1993) classification, on the one hand, we examine a set of verbs of motion belonging to the subclass of "run" verbs, which are embedded within the group of manner of motion verbs. Particularly, we deal with these three "run" verbs: *trek*, *glide*, and *hike*. On the other hand, also following Levin's (1993) typology, we examine a set of verbs of motion belonging to the subclass of verbs that are vehicle names, which are embedded within the group of verbs of motion using a vehicle. Specifically, we analyse these three verbs: *raft*, *canoe*, and *parachute*.

In doing so, this study follows a usage-based approach and aims at contributing to the syntactic and semantic characterisation of real motion verbs in the specialised domain of adventure tourism, paying particular attention to the structural and relational link between these verbs and the behaviour of their participants in terms of argument structure realisation, and focusing on the different syntactic alternations in which these verbs can be found, according to Levin's (1993) ideas. The findings of this study are expected to provide some insights into the linguistic features of the specialised language of adventure tourism in English, as its lexico-semantic and syntactic behaviour is more restricted than its counterpart in general language. Moreover, they could be implemented to further elaborate the characterisation of this type of verbs and their argument structure in tools that work with this specialised language, such as *DicoAdventure*, the online dictionary about adventure tourism.[1]

This chapter is organised as follows. Section 2 provides an overview of motion verbs and their argument structure realisation, mainly focusing on the different syntactic alternations in which these verbs can be found. Section 3 describes the methodology employed and the steps followed in the processes of extraction of data and classification of the instances retrieved. After that, Section 4 presents the main findings of the study and provides a discussion of the results. Finally, Section 5 offers some final remarks and sketches future lines of research.

2. The Motion Event and its Argument Structure

In a motion event, different entities participate and relate to each other in diverse ways, depending on the argument structure realisation of the verb. Therefore, syntactically distinct but semantically related alternations can be found in

---

1 http://olst.ling.umontreal.ca/dicoadventure/ [Last accessed: 28/09/2023].

discourse. For example, some constructions may include animate or inanimate entities as subject, they may incorporate direct objects in their transitive forms or oblique structures in their intransitive counterparts, some may encompass prepositional phrases or preposition drop alternations, and the like.

At a syntactic and lexico-semantic level of analysis, a number of scholars have contributed to the depiction of the different entities that participate in the motion event. For instance, Talmy (2000, p. 311) uses the terms *Figure* and *Ground* to refer to the entity that moves in space (Figure) with respect to another stationary entity (Ground). The label Figure refers to both animate and inanimate entities appearing in, respectively, real and fictive motion events. In the basic or unmarked syntactic form of a non-agentive clause, the semantic role of Figure coincides with the grammatical subject, whereas the notion of Ground is identified with the grammatical (oblique) object. In turn, in the basic or unmarked syntactic form of an agentive clause, where the grammatical subject is the Agent, the notion of Figure coincides with the grammatical object and the Ground has an oblique object function. As Talmy (2000) puts it, "[a]ny Figure/Ground assignments other than these are taken to be nonbasic or derived" (p. 334). Such non-basic or marked Figure/Ground assignments correspond to different syntactic patterns, which Levin (1993) names syntactic alternations, as detailed in Section 2.1. Such syntactic alternations also imply semantic and pragmatic distinctions with respect to the basic or unmarked clause from which they are lexically and/or syntactically derived.

In addition, Talmy (2000) uses the terms *Path* and *Motion* to refer to other types of entities that can participate in the motion event. The term Path refers to the path followed or occupied by the Figure, whereas the term Motion involves the idea of motion itself or locatedness denoted in the event. Other scholars (Levin & Rappaport Hovav, 1992; Mani & Pustejovsky, 2012) have also proposed other entities within the motion event, such as *Source* and *Goal*, *Manner*, and *Direction*. The Source is defined as the location where the motion is initiated, whereas the Goal is identified with the location where the motion terminates or is directed to. The notion of Manner refers to the way in which the motion is carried out, and the notion of Direction involves the direction itself that is followed in the motion event.

In the specialised language of adventure tourism, Durán-Muñoz and L'Homme (2020, p. 50) propose a detailed analysis of the argument structure of verbs of motion, identifying arguments and circumstantials. The former involve more central elements (namely, arguments, like TOURIST, PLACE, DIRECTION, SOURCE, DESTINATION, and PATH), whereas the latter suggest more peripheral or optional elements (like MANNER, DISTANCE, DURATION, and FREQUENCY). Both arguments and circumstantials enrich the lexico-semantic structural relations of the motion verbs they accompany, as they specify their meaning by activating the specialised knowledge of these motion verbs in context. In other words, "the semantic information of the complement (of a verb) contributes to the specification of a unique and appropriate meaning of the verb" (Yoshimura, 1998, p. 114). This idea is fundamental to better understand the inherent features of a specialised domain, as it is the language of adventure tourism, and studies like the present one aim at contributing to this end.

2.1. Syntactic alternations with verbs of motion

According to Levin (1993, p. 263), the class of motion verbs involves a wide range of predicates, including the following: (i) verbs of inherently directed motion (e.g., *ascend*); (ii) the class of "leave" verbs (e.g., *abandon*); (iii) manner of motion verbs, divided into the subclass of "roll" verbs (e.g., *slide*), on the one hand, and the subclass of "run" verbs (e.g., *jump*), on the other; (iv) verbs of motion using a vehicle, split into the subclass of verbs that are vehicle names (e.g., *kayak*), on the one hand, and the subclass of verbs that are not vehicle names (e.g., *sail*), on the other; (v) the class of "waltz" verbs (e.g., *dance*); (vi) the class of "chase" verbs (e.g., *trail*); and (vii) the class of "accompany" verbs (e.g., *guide*).

As it is explained in Section 3, this study focuses on the analysis of a set of motion verbs belonging to the abovementioned classes of manner of motion verbs (especially, the subclass of "run" verbs) and verbs of motion using a vehicle (more specifically, the subclass of verbs that are vehicle names). These two classes of verbs have been chosen due to their productivity and relevance in the domain of adventure tourism, as demonstrated in Durán-Muñoz and Jiménez-Navarro (2023). For this reason, in Sections 2.1.1. and 2.1.2. we examine only those syntactic alternations that involve these two classes of verbs, following Levin's (1993) proposal.

2.1.1. *The subclass of "run" verbs and their syntactic alternations*

As Levin (1993, p. 264) explains, the class of manner of motion verbs encompasses a type of motion that is frequently (though not necessarily) associated with displacement, without specifying an inherent direction at a semantic level of analysis. These verbs are related to a notion of manner or means of motion in their meaning, and they differ in the particular manner or means dealt with in the motion event. This class of verbs is divided into two subclasses: the so-called "roll" verbs and the subclass of "run" verbs; in this study, we focus on the latter. The "run" verbs selected here are *glide*, *hike*, and *trek*. The subclass of "run" verbs is characterised by describing the manner in which an animate entity can move; thus, they describe the displacement of an entity in a particular manner or by a particular means. Besides, they do not specify the direction of motion, unless they appear with an explicit directional phrase (Levin, 1993, p. 267). The subclass of "run" verbs can be found in the following syntactic alternations,[2] according to Levin (1993, p. 266): (1) the basic or unmarked clause, (2) the induced action alternation, (3) the locative preposition drop alternation, (4) the *there*-insertion alternation, (5) the locative inversion, (6) the measure phrase alternation, and (7) the resultative phrase alternation. In addition, when combining Levin's (1993) analysis and a lexico-semantic approach to argument structure (cf. Talmy, 2000), we can characterise these syntactic alternations as follows:

1. The basic or unmarked clause, as in *The horse jumped over/across/into/out of the stream*. In the basic or unmarked clause, an animate entity is profiled as the grammatical subject and has the semantic role of Figure, whereas the entity that receives secondary focal prominence occurs in object position and is embedded within a prepositional phrase, fulfilling the semantic role of Ground.[3]
2. The induced action alternation, as in *Tom jumped the horse over the fence*. In the induced action alternation, three arguments participate in the clause, being the first two animate entities and the last one inanimate: (1) the Agent (promoted to subjecthood), (2) the Patient (occupying object position and fulfilling the semantic role of Figure), and (3) the Oblique participant (embedded within a prepositional phrase and serving as the Ground).
3. The locative preposition drop alternation, as in *The horse jumped the stream*. In the case of the so-called locative preposition drop alternation, we find a

---

2 Levin (1993, pp. 266–267) also presents other alternations that have a lexical rather than syntactic nature within the subclass of "run" verbs (namely, the so-called "adjectival passive participle" alternation, the "adjectival perfect participle" alternation, the "cognate object" alternation, and the "zero-related nominal" alternation). However, these lexical alternations are not examined in this paper because the purpose of this study is to analyse the different syntactic alternations provided by the set of motion verbs chosen.

3 The role of Ground in Talmy's (2000) theory is further detailed in Durán-Muñoz and L'Homme (2020) in order to capture the intricate nature of the specialised language of adventure tourism. The authors specify the lexico-semantic subtype of Ground depending on the features of the prepositional (or nominal) phrase in question; thus, they distinguish among PATH, DESTINATION, PLACE, SOURCE, DESTINATION, among others. In this paper, this fine-grained distinction at the lexico-semantic level is not carried out, as we use the umbrella term Ground to refer to both object and oblique object participants.

Figure-Ground configuration which is identical to the one in the basic or unmarked case (number 1. above), but now the object is not an oblique participant, as it is not introduced by any preposition. The participant that fulfils the role of Figure is an animate entity, whereas the second argument is inanimate, as it still retains a locative value at a semantic level of analysis.

4. The *there*-insertion alternation, as in *There jumped out of the box a little white rabbit*. The *there*-insertion alternation presents a particular syntactic pattern which is more frequent in tales and literature than in other domains of language. Whereas the basic form displays a Figure-Ground configuration in which the Figure is profiled and the Ground is defocused, in the *there*-insertion alternation the order of arguments is reversed (Ground-Figure), and they are preceded by the inserted adverbial *there*, followed by the motion verb in question. The Ground still occurs as an oblique participant in this alternation.

5. The locative inversion, as in *Out of the box jumped a little white rabbit*. The locative inversion alternation also reverses the prototypical configuration Figure-Ground, so it profiles the Ground over the defocused Figure. In this occasion, the Ground still appears as an oblique participant (embedded within a prepositional phrase), but no other elements precede it in sentence structure (as it happens in the *there*-insertion alternation).

6. The measure phrase alternation, as in *We walked five miles*. The so-called measure phrase alternation is syntactically identical to the locative preposition drop alternation (number 3. above), but they are semantically different. That is, in both cases, the Figure-Ground configuration is the same, but they differ from the basic or unmarked form (number 1. above) in that the Ground is found in object (not oblique object) position (i.e., without being embedded in a prepositional phrase). However, in the measure phrase alternation, the Ground is semantically limited to the referencing of measurement entities in terms of space (miles, kilometres, etc.) or time (minutes, hours, days, etc.), among others (Corver, 2009).

7. The resultative phrase alternation, as in *Tom ran the soles off his shoes*. Finally, the resultative phrase alternation presents the Ground entity in a way that it highlights the resultative facet of the interaction between the entity and the motion event. As the author (Levin, 1993) defines it, "[a] resultative phrase is an XP which describes the state achieved by the referent of the noun phrase it is predicated of as a result of the action named by the verb" (p. 101). In the resultative alternation, the Agent (Figure) is promoted to subjecthood and the Patient (Ground) receives secondary focal prominence. The patientive (Ground) entity contains "a resultative element that may appear in the form

of an adverbial/adjectival/participial/infinitival complementation which adds an emphatic value to the construction" (Palma Gutiérrez, 2024/forthcoming).

2.1.2. *The subclass of verbs that are vehicle names and their syntactic alternations*

As Levin (1993) explains, this class of verbs of motion are all "zero-related to nouns that are vehicle names", that is, they basically mean "go using the vehicle named by the noun" (p. 268). This type of verbs describes the motion of an entity, but, like the subclass of "run" verbs, these verbs with vehicle names do not imply a specific direction of motion unless there is an explicit directional phrase to do so. This type of verbs can be found in the following syntactic alternations,[4] according to Levin (1993, pp. 267–268):

1. The basic or unmarked form, as in *They skated*.
2. The intransitive form with a locative phrase, as in *They skated along the canal*.
3. The induced action alternation, as in *He skated Penny around the rink*.
4. The locative preposition drop alternation, as in *They skated the canals*.
5. The resultative phrase alternation, as in *Penny skated her skate blades blunt*.

In the basic or unmarked form (number 1. above), there is only one explicit entity, the Agent, which is logically promoted to subjecthood. This participant fulfils the semantic role of Figure, whereas the role of the Ground is implicit in discourse though not syntactically coded.

In the case of the intransitive form which contains a locative phrase (number 2. above), we find exactly the same Figure-Ground configuration as in the basic or unmarked form of the subclass of "run" verbs dealt with in Section 2.1.1. That is, the animate agentive participant occurs in subject position, being identified with the semantic role of Figure. In turn, the role of Ground is carried out by an inanimate entity that receives secondary focal prominence and is, thus, found in oblique object position, as it is embedded within a prepositional phrase.

The main difference between the subclass of motion verbs that are vehicle names and the subclass of "run" verbs, according to Levin (1993), is that the default (basic or unmarked) form in the former is a one-argument structure, whereas it is a two-argument structure (with a prepositional locative phrase)

---

4   Levin (1993, p. 268) mentions that, in her study, the subclass of verbs of motion that are vehicle names has not been attested in the *there*-insertion or the locative inversion alternations, though it could be possible to find them in these constructions. In the present corpus-driven study, we have not found any cases.

in the latter. As it is discussed in Section 4, this idea might be challenged by the results obtained in the corpus analysis carried out here, since they highlight the productivity of the syntactic structures which contain prepositional locative phrases in both cases.

The rest of alternations in which the subclass of motion verbs that are vehicle names can be found – i.e., the induced action alternation (number 3. above), the locative preposition drop alternation (number 4. above), and the resultative phrase alternation (number 5. above) – have already been explained in Section 2.1.1., as their syntactic and semantic behaviour is the same as in the subclass of "run" verbs.

Choosing either the basic, unmarked form or a non-basic, syntactically derived alternation is a matter of entity-profiling/defocusing phenomenon which is conducted by a particular speaker at a given moment. Therefore, the speaker's selection process implies syntactic, semantic, and discourse-pragmatic differences. As Palma Gutiérrez (2024/forthcoming) points out, "[i]n the cognitive schema of an event, the speaker has different options to codify their message. The speaker's actual choice of any of the potential options is however determined by the contextual situation". This idea is of particular relevance in the analysis of a specialised discourse, as it leads to the examination of the intrinsic features of verbs and their argument structure realisation in comparison to their counterparts in general language. The present study analyses the lexico-semantic and syntactic restrictions upon the specialised discourse of adventure tourism in terms of its most productive verbs, their syntactic patterns, and their argument structure realisation in real usage, as compared to a less constrained use of these verbs and their argument structure in general language.

3. Methodology

This chapter is based on a corpus study of contextualised examples retrieved from the specialised corpus ADVENCOR, which is a bilingual (English-Spanish) specialised corpus made of English and Spanish promotional texts that are representative of the domain of adventure tourism. It has been used in previous studies with terminological and phraseological purposes (cf. Durán-Muñoz & Jiménez-Navarro, 2023; Durán-Muñoz & L'Homme, 2020, among others), and it is now employed to analyse the syntax-semantics interface of the verbs of motion selected in the present study. In order to explain why we have chosen this particular set of six motion verbs (namely, *trek*, *glide*, *hike*, *raft*, *canoe*, and *parachute*), we need to look at the process of extraction and selection of verbs of motion as carried out in previous research studies; our decision is based on this process.

In Durán-Muñoz and Jiménez-Navarro (2023), the authors use the *Keywords* function of Sketch Engine[5] to automatically extract the verbal units. The relevant aspect here is that the *Keywords* function uses *simple math* (Kilgarriff, 2009) as keyness score. In doing so, the frequency and relative frequency of the same words are compared in both the specialised corpus (ADVENCOR) and the reference corpus chosen (in this case, the enTenTen20 corpus, now updated as the enTenTen21 corpus, since it was the densest available corpus in English and the one set by default). The results of this extraction process showed the candidate terms that were specific regarding the specialised corpus, which proved their productivity and relevance in the domain of adventure tourism. After a process of manual pruning (for instance, it included the discardment of wrongly tagged verbs as well as of wrongly lemmatised units), a final list of 152 motion verbs was selected for the analysis undertaken in Durán-Muñoz and Jiménez-Navarro (2023).

Having said that, the methodology of the present study starts from the selection of the top-10 motion verbs analysed in Durán-Muñoz and Jiménez-Navarro (2023). These top-10 verbs were ordered with regard to their keyness score: *skydive, raft, trek, rappel, abseil, canoe, glide, hike, parachute,* and *mountaineer.* However, four of them were discarded because they were not classified in Levin's (1993) typology of verbs of motion, and, thus, no syntactic alternations were provided for them in this author's work. The discarded verbs were *skydive, rappel, abseil,* and *mountaineer.* Table 1 illustrates the verbs finally selected for this study, ordered according to their keyness score.

Table 1. Selected motion verbs ordered according to their keyness score

Motion verb	Keyness score
raft	314.327
trek	244.583
canoe	109.166
glide	89.336
hike	81.650
parachute	79.965

After the selection process, the following step consisted in identifying their subtypes of verbs of motion according to Levin's (1993) classification and they were divided into two subclasses. On the one hand, *trek, glide,* and *hike* are found in Levin's (1993, p. 265) typology as members of the subclass of "run" verbs

---

5   https://www.sketchengine.eu/ [Last accessed: 28/09/2023].

(embedded within the class of manner of motion verbs). On the other hand, *raft*, *canoe*, and *parachute* are classified by Levin (1993, p. 267) as members of the subclass of verbs that are vehicle names (embedded within the class of verbs of motion using a vehicle).

In order to carry out the retrieval of instances in context, the *Concordance* function of Sketch Engine was used. To do so, the advanced options "lemma" and "verb" were selected and applied in each of the six selected verbs. A manual process to discard the non-valid instances was implemented by exhaustively examining all the instances found one by one, since the query system of Sketch Engine does not allow automatic specifications that filter out lexical and/or morphological mismatches in context; therefore, the following cases were not considered:

1. Nominalised forms of verbs ending in *-ing* (e.g., *trekking*-n).
2. Zero-related nominals, that is, nouns wrongly tagged as verbs by the term extractor (e.g., *treks*-n).
3. Adjectives lemmatised as verbs (e.g., *trekking*-j, *trekked*-j).
4. Absolute intransitive forms with the subclass of "run" verbs, that is, one-argument structures with just a syntactically coded subject but no locative oblique object (e.g., *You can also hike with the bungee America group by paying $10 fees*).

Table 2 includes the six motion verbs selected for this study, ordered according to their frequency of occurrence in the corpus examined; the final number of valid instances retrieved in each case after a manual discarding process is also included.

Table 2. Selected motion verbs ordered according to their frequency of occurrence and number of valid instances retrieved

Motion verb	Frequency of occurrence	Number of valid instances
hike	1,446	276
trek	1,254	73
raft	1,086	128
glide	471	62
canoe	332	30
parachute	152	9

After the discarding process, the total number of valid instances were classified according to the syntactic alternations they represented in relation to the type of verb of motion and the argument structure displayed in each case. In Section 4 the most relevant results are discussed.

4. Results and Discussion

The total number of corpus contexts examined by using the *Concordance* function of Sketch Engine was 4,741 instances, and, after the discarding process detailed in the previous section, the total number of valid instances for our analysis was 578 (i.e., 12.2 % of the contexts), out of which 411 contextualised examples incorporated a predicate of the subclass of "run" verbs and the remaining 167 instances belonged to the subclass of motion verbs that are vehicle names. In the following paragraphs, we analyse the 578 occurrences in terms of the number of tokens that participate with each motion verb and the syntactic alternation found. Figure 1 unpacks the number of instances found with regard to the three motion verbs that belong to the subclass of "run" verbs (namely, *trek*, *glide*, and *hike*), whereas Figure 2 does the same regarding the subclass of motion verbs that are vehicle names (specifically, *raft*, *canoe*, and *parachute*). In both Figure 1 and Figure 2 special attention is paid to the range of syntactic alternations coded in each case.

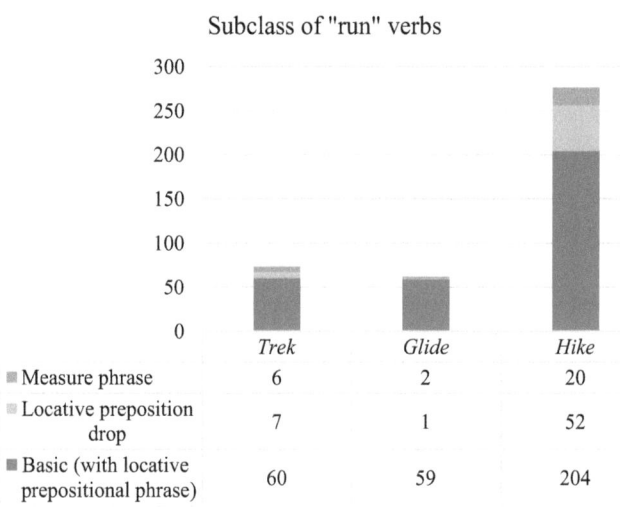

	Trek	Glide	Hike
Measure phrase	6	2	20
Locative preposition drop	7	1	52
Basic (with locative prepositional phrase)	60	59	204

Figure 1. "Run" verbs and their syntactic alternations

As can be observed in Figure 1, the verb *hike* was much more productive in ADVENCOR than the other two "run" verbs, *trek* and *glide*. The three of them have in common a higher frequency of occurrence in their basic or unmarked form incorporating a locative prepositional phrase than in the rest of alternations. Nevertheless, the productivity of other syntactic alternations is remarkable in the case of the verb *hike*.

Figure 1 only focuses on three out of the seven syntactic alternations proposed by Levin (1993) with regard to the subclass of "run" verbs, as there were no cases of the other alternations, that is, the induced action alternation, the *there*-insertion alternation, the locative inversion alternation, and the resultative phrase alternation. The lack of instances constructed with these syntactic alternations leads to a better understanding of the specialised domain of adventure tourism, as it indicates their low productivity in the field of study compared to the basic or unmarked form which incorporates a locative prepositional phrase (example (1)), the locative preposition drop alternation (example (2)), and the measure phrase alternation (example (3)).

(1) You will *glide* through the trees and encounter a world unknown.[6]
(2) You *are trekking* the same route but here we change the path.
(3) We *hiked* another few miles exploring […].

In Figure 2 we present the verbs of motion that are vehicle names and their distribution in terms of the syntactic alternations with which they were found in the corpus consulted.

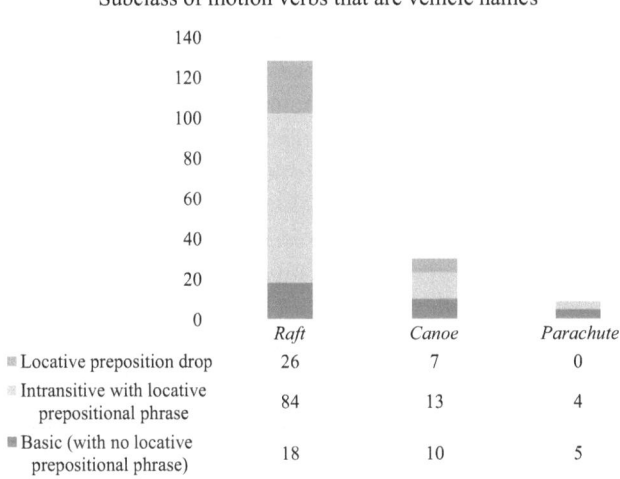

Figure 2. Verbs of motion that are vehicle names and their syntactic alternations

---

6   These examples were retrieved from ADVENCOR. Italics have been added to highlight the verbs under study.

As illustrated in Figure 2, in the vehicle names subclass the verb *raft* was much more productive in terms of the variety of syntactic alternations realised than the other two motion verbs, that is, *canoe* and *parachute*. In contrast to the results obtained in the subclass of "run" verbs, the verbs *raft* and *canoe* are more productive in other than their basic or unmarked form. In the case of the motion verbs that are vehicle names, Levin (1993) suggests that the basic form is the absolute intransitive construction (i.e., a one-argument structure) based on the premise that this form is more frequent than the rest of alternations proposed (p. 267). However, as shown in Figure 2, when used in the specialised domain of adventure tourism, *raft* and *canoe* show a tendency to occur more productively in two-argument structures (those intransitives that incorporate a locative prepositional phrase to introduce the oblique argument) (84 tokens and 13 tokens, respectively). Besides, in the case of *raft*, the number of instances following a locative preposition drop pattern is also superior to the number of examples found in the basic or unmarked form (26 vs 18, respectively).

Figure 2 only focuses on three out of the five syntactic alternations proposed by Levin (1993) in relation to the subclass of motion verbs that are vehicle names, as there were no cases of the other alternations, that is, the induced action alternation and the resultative phrase alternation. As discussed in the case of the subclass of "run" verbs, the lack of examples incorporating these syntactic alternations in the corpus under study leads, in fact, to the identification of a more restricted and strict linguistic behaviour of these verbs and their argument structure in the specialised language of adventure tourism, compared to their counterparts in general language. Thus, the three alternations shown in Figure 2 are the basic or unmarked form which is identified with an absolute intransitive that does not incorporate any locative prepositional phrase (example (4)), the intransitive form which incorporates a locative prepositional phrase to introduce an oblique argument (example (5)), and the locative preposition drop alternation (example (6)).

(4) We *had* never *rafted* before.
(5) Russian skydiver Valery Rozov *parachuted* into an active volcano here.
(6) *Canoe* the River Wye.[7]

---

7   Example (6) incorporates the verb *canoe* in its imperative form; however, it is included here because, at the semantic level of analysis, the imperative form implies an agentive participant ("you").

5. Conclusions

This chapter aimed at contributing to the characterisation of the syntactic and semantic constraints of the specialised discourse of adventure tourism in English by analysing the linguistic behaviour and frequency of occurrence of motion verbs through a corpus-driven study. Particularly, we focused on the examination of the syntactic alternations in which a set of six motion verbs can be found in context, namely, *trek*, *glide*, *hike*, *raft*, *canoe*, and *parachute*. The sample of contextualised instances that were analysed here was extracted from the specialised corpus on adventure tourism ADVENCOR.

The objectives set in this usage-based study were the following: first, to provide an overview of the motion event and its argument structure realisation, pointing at the case of the specialised domain of adventure tourism, and second, to analyse the structural and relational link between the six motion verbs eventually chosen and the behaviour of their corresponding participants in terms of argument structure realisation, paying special attention to the different syntactic alternations in which these verbs can be found, according to Levin's (1993) ideas. As a result, this corpus-driven work provides some insights into the linguistic features of the specialised language of adventure tourism in English.

A sample of 4,741 instances was examined, and, after a discarding process to eliminate those cases that did not meet the requirements set (over 85 % of them), the total number of valid instances for our analysis was 578. Once the final sample was analysed, the contextualised instances were first classified in terms of the subclass of motion verbs they belonged to (as proposed by Levin, 1993) and, then, in terms of the syntactic alternations displayed in every case, paying attention to the argument structure realised. In the case of the subclass of "run" verbs, the motion verb *hike* was more productive than the other two motion verbs examined here (namely, *trek* and *glide*), both in terms of the number of total instances and the number of non-basic or derived alternations found. Similarly, among the subclass of motion verbs that are vehicle names, the predicate *raft* was more frequent than the other two motion verbs (i.e., *canoe* and *parachute*), and it was also more prominent in terms of the number of instances involving non-basic alternations. Remarkably, the data examined reveals that, in the cases of *raft* and *canoe*, the most productive structures were not the basic forms but the intransitive structures which incorporate oblique arguments.

All in all, the objectives set in this study were achieved, as they attempted to contribute to the characterisation of the syntactic and semantic features of the motion verbs selected and their argument structure realisation in the specialised discourse of adventure tourism. The main findings of this study demonstrate that the motion verbs analysed and their argument structures exhibit a more limited, restricting, and strict behaviour when found in the field under study, as compared to their counterparts in general language. This is so because some of the syntactic alternations addressed by Levin (1993) were not detected in Advencor, denoting a more constrained nature of these grammatical structures in the specialised domain. Hopefully, this analysis will encourage further research on motion verbs. For instance, a possible future extension of the present study could examine the syntactic and lexico-semantic behaviour of other prominent motion verbs in the specialised domain of adventure tourism, like those predicates discarded here for not being included in Levin's (1993) classification (namely, *skydive*, *rappel*, *abseil*, and *mountaineer*).

References

Buendía-Castro, M. (2021). Verb dynamics. *Terminology*, *18*(2), 149–166. https://doi.org/10.1075/term.18.2.01bue

Buendía-Castro, M. (This volume). Lexical domains in the field of adventure tourism. In I. Durán-Muñoz & E. L. Jiménez-Navarro (Eds.), *Exploring the language of adventure tourism: A corpus-assisted approach*. Peter Lang.

Casademont, A. J. (2014). On the elements activating the transmission of specialized knowledge in verbs. *Terminology*, *20*(1), 92–116. https://doi.org/10.1075/term.20.1.05joa

Corver, N. (2009). Getting the (syntactic) measure of measure phrases. *The Linguistic Review*, *26*(1), 67–134. https://doi.org/10.1515/tlir.2009.003

Durán-Muñoz, I., & Jiménez-Navarro, E. L. (2023). Motion verbs in adventure tourism: A lexico-semantic approach to fictive meaning. *IJES*, *23*(1), 27–48. https://doi.org/10.6018/ijes.532851

Durán-Muñoz, I., & Jiménez-Navarro, E. L. (2021). Colocaciones verbales en el turismo de aventura: Estudio contrastivo inglés-español. In G. Corpas Pastor, M.ª R. Bautista Zambrana, & C. M. Hidalgo-Ternero (Eds.), *Sistemas fraseológicos en contraste: Enfoques computacionales y de corpus* (pp. 121–142). Comares.

Durán-Muñoz, I., & L'Homme, M.-C. (2020). Diving into adventure tourism from a lexico-semantic approach: An analysis of English motion verbs. *Terminology*, *26*(1), 33–59. https://doi.org/10.1075/term.00041.dur

Fellbaum, C. (1990). English verbs as a semantic net. *International Journal of Lexicography*, *3*(4), 278–301. https://doi.org/10.1093/ijl/3.4.278

Hale, K., & Keyser, S. J. (1998). The basic elements of argument structure. In H. Harley (Ed.), *Papers from the UPenn/MIT Roundtable on Argument Structure and Aspect* (pp. 1–47). The MIT Press.

Jacinto García, E. (This volume). The argument structure of motion verbs in Spanish: A methodological proposal applied to *DicoAdventure*. In I. Durán-Muñoz & E. L. Jiménez-Navarro (Eds.), *Exploring the language of adventure tourism: A corpus-assisted approach*. Peter Lang.

Kilgarriff, A. (2009). Simple maths for keywords. In M. Mahlberg, V. González-Díaz, & C. Smith (Eds.), *Proceedings of the Corpus Linguistics Conference 2009 (CL2009)* (pp. 1–6). University of Liverpool.

Langacker, R. W. (1986). Abstract motion. In V. Nikiforidou, M. VanClay, M. Niepokuj, & D. Feder (Eds.), *Proceedings of the Twelfth Annual Meeting of Berkeley Linguistics Society* (pp. 455–471). Berkeley Linguistics Society.

Levin, B. (1993). *English verb classes and alternations. A preliminary investigation.* The University of Chicago Press.

Levin, B., & Rappaport Hovav, M. (1992). The lexical semantics of verbs of motion. The perspective from unaccusativity. In I. M. Roca (Ed.), *Thematic structure: Its role in grammar* (pp. 247–269). De Gruyter Mouton. https://doi.org/10.1515/9783110872613.247

L'Homme, M.-C. (1998). Le statut du verbe en langue de spécialité et sa description lexicographique. *Cahiers de Lexicographie*, *73*(2), 61–84.

L'Homme, M.-C. (This volume). Frame Semantics and domain-specific resources. In I. Durán-Muñoz & E. L. Jiménez-Navarro (Eds.), *Exploring the language of adventure tourism: A corpus-assisted approach*. Peter Lang.

Mani, I., & Pustejovsky, J. (2012). *Interpreting motion: Grounded representations for spatial language*. Oxford University Press.

Matlock, T. (2004). The conceptual motivation of fictive motion. In G. Radden & K.-U. Panther (Eds.), *Studies in linguistic motivation* (pp. 221–248). De Gruyter Mouton.

Palma Gutiérrez, M. (2024/forthcoming). Profiling and defocusing phenomena in the discourse of fe/male novelists: A corpus-based approach. In E. L. Jiménez-Navarro & L. M. Martínez Serrano (Eds.), *Where gender and corpora meet: New insights into discourse analysis*. Peter Lang.

Sager, J. C. (1990). *A practical course in terminology processing*. John Benjamins Publishing Company.

Talmy, L. (2000). *Toward a cognitive semantics, Vol. 1. Concept structuring systems*. The MIT Press.

Vargas Sierra, C. (2012). La tecnología de corpus en el contexto profesional y académico de la traducción y la terminología: Panorama actual, recursos y perspectivas. In M. A. Candel Mora & E. Ortega Arjonilla (Eds.), *Tecnología, traducción y cultura* (pp. 67–99). Tirant Humanidades.

Yoshimura, K. (1998). Encyclopedic structure in nominals and middle expressions in English. *Kobe Papers in Linguistics, 1*, 112–140.

Acknowledgements

A special thanks goes to my department mates, Isabel Durán-Muñoz and Eva Lucía Jiménez-Navarro, for their invaluable help and fellowship, and for giving me the opportunity to participate in this volume.

Carmen Portero Muñoz

Universidad de Córdoba

ff1pomuc@uco.es

10 The use of compounds in the adventure tourism lexicon

Abstract: This chapter explores the relevance of English compounding processes in the lexical repertoire of adventure tourism. The focus will be on compound nouns, specifically those with a morphologically complex head noun ending in *-ing*, such as *mountain biking*, typically used to designate different adventure activities. By using corpus data, the different compound patterns referring to adventure tourism activities and their relative productivity will be identified. First, compounds will be classified syntactically, and then they will be analysed in terms of the denotation of the first word, the head noun, and the internal association between them, that is, thematic relations. It will be shown that compounding is a significant morphological process for the formation of new words in the realm of this tourism segment. In addition, it will be argued that adverbial-type roles (e.g., PATH, PLACE, SOURCE) become central in meaning relations expressed by noun compounds with a verb-derived head within the specific area of adventure tourism.

Keywords: ADVENCOR corpus, adventure tourism, compounding, metonymy, synthetic compound, thematic role

1. Introduction

The present study is an exploration of the lexicon of a unique semantic domain: adventure tourism. More specifically, the focus is on one specific type of word formation process whereby words are created in this realm, namely, compounding, one of the most productive processes that are used in English to create new words (Plag, 2003, p. 132).

One of the factors that triggers the coinage of new lexical items is the need for referring to new aspects of the real world, such as inventions, commercial items, political, or sanitary circumstances. In the case of adventure tourism, the emergence of new activities boosts the demand for new verbs to designate them. Motion verbs are thus particularly relevant in this specific area as they are used to express different actions in which a number of participants, such as tourists, instruments, or places are involved (Durán-Muñoz & L'Homme, 2020). It is not surprising then that a variety of nouns also emerge to refer to the activities associated with the different actions, many of which are compound nouns.

The purpose of the present study is twofold. The first and more general aim will be to make a small contribution to current research on English compounding, showing the relevance of this morphological process in the coinage of nouns for adventure tourism activities. Second, and more specifically, this survey is intended as an exploration of the specific formal and semantic properties of the different compounds, as well as of the relative productivity of the patterns found. This is a multidisciplinary piece of research which provides a clear illustration of the way in which linguistic analysis, and particularly morphological theory and description, can help identify crucial information about the use of English for specific purposes, the terminology of adventure tourism.

This chapter is structured as follows. Compounding is first defined, and a brief outline of the different syntactic and semantic types that have been drawn in the morphological tradition is presented in Section 2. This is followed by a description of methodological issues in Section 3. Section 4 furnishes a syntactic and semantic analysis of the corpus data and discusses the results. Finally, the main findings and implications of the analysis are briefly summarised in a conclusion in Section 5.

2. Compounding

Compounds have been defined as "a lexical unit made up of two or more elements, each of which can function as a lexeme independent of the other(s) in other contexts, and which show some phonological and/or grammatical isolation from normal syntactic usage" (Bauer, 2001, p. 695). In spite of this apparently neat definition, compounds are troublesome linguistic units. For example, the decision on the status of a two-word sequence as a morphological unit, that is, a compound, or as a syntactic unit, that is, a phrase, is not always easy to make.

In the morphological tradition, scholars have proposed criteria of different types in an attempt to identify compounds (Bauer et al., 2013; Matthews, 1974; Plag, 2003). One of these criteria is orthographical conventions, which is not a comprehensive guide, as new compounds are usually written with open spelling, that is, as two separate words, in which case they look like phrases. A second important criterion that is often alluded to is stress, which is regarded to fall on the first word in compounds, as compared to phrases. This has nevertheless been shown to fail in numerous cases and is not reliable, especially in the case of verb compounds and adjective compounds (Plag, 2003, 2006). Compounds have also been distinguished as semantically unpredictable units (e.g., *blackmail*), or units where the meaning of the component words is more specific in the compound than outside it (e.g., *greenhouse*). However, compounds can also be transparent, in other words, they can have predictable meanings (e.g., *headache*). Finally, the syntactic potential of compounds is more limited than that of free collocations. Thus, it is traditionally argued that they cannot be interrupted by another word (e.g., **a library boring book*) neither expanded by coordinating the head noun with other units (e.g., **a heart and Thai massage*). In the case of Adjective–Noun compounds, it has been stated that the adjective cannot be premodified independently (e.g., **a very greenhouse*).

All the previous criteria are relevant and usually helpful in the case of lexicalised compounds. Nevertheless, none of them is sufficient, and it is often the case that different criteria are in conflict when trying to decide on specific compounds (Bauer, 1998). It can therefore be concluded that compounds are not a strict category and that there is a large number of sequences of two (or even more) words that are compound-like, despite not being well established yet.

2.1. Classification of English compounds

This section introduces various categories of compounds that have been established in the morphological tradition from semantic and syntactic perspectives.

2.1.1. Semantic classification

English compounds are classified into three main semantic types depending on where the head of the compound is found. The head is the element that determines the grammatical category of the compound, which is, in most cases, the right-hand side component in the case of English. Semantics-wise, the head is the word designating the entity that the whole compound refers to.

These three categories are the following. First, endocentric compounds are those that can be reduced to the head because they denote a more specific type of the entity represented by the head. For example, *air-frying* is endocentric, since it denotes the action of frying using air instead of oil; thus, it is a different modality of frying. On the other hand, exocentric compounds are those that refer to an external entity, that is, the semantic head of the compound is none of its components. For instance, *housewarming* does not express a more specific type of warming, but "a party that you give when you move into a new house" (Cambridge Dictionary, n.d.). Finally, coordinative compounds are those whose components have equal semantic weight and that are hyponyms of each of these components. As an illustration, a *comedy-thriller* is both a subtype of comedy and a more specific type of thriller.

2.1.2. Syntactic classification

In the morphological tradition, compounds have been analysed syntactically as well as semantically. A first syntactic distinction is made according to the syntactic category or word class the compound belongs to and, therefore, there are noun compounds, adjective compounds, and verb compounds. Different subtypes are then distinguished in terms of their internal structure, that is, the word class of the component words, as well as the syntactic-semantic relation between them. As regards the word class of the compound's components, English offers a wide array of patterns, which include noun, adjective, and verb compounds with a first nominal (N), adjectival (A), prepositional (P), or verbal (V) constituent, as shown in Table 1, but they differ in productivity. Thus, verb compounds are not very common in English.

Table 1. Syntactic classification of English compounds

Noun compounds	NN	*apron string*	AN	*high school*	PN	*overdose*	VN	*swearword*
Adjective compounds	NA	*headstrong*	AA	*icy cold*	PA	*overwide*	VA	*fail-safe*
Verb compounds	NV	*brainwash*	AV	*blackmail*	PV	*outlive*	VV	*stir-fry*

Finally, on the grounds that they have structures that are syntax-like, compounds have been analysed in terms of different syntactic or semantic relations that connect the component words (Bauer et al., 2013; Matthews, 1974; Quirk et al., 1985). This means that basic clause relations have been transposed to morphological structure. For example, in *poetry writer*, *poetry* is the result of writing and *writer* is the agent of the action, which, translated into syntactic analysis, means that the first noun is a kind of object of the second noun. By contrast, in *swimming-pool*, *pool* is the place where the swimming occurs. In this case, the second noun can be analysed syntactically as a kind of adverbial modification (expressing place) of the first noun.

In order to account for the multiple relations within compounds, scholars have presented different proposals (Adams, 1973, 2001; Bauer, 1983, 2001; Bauer et al., 2013; Bisetto & Scalise, 2005; Quirk et al., 1985; Selkirk, 1982), although most of them agree on a twofold distinction between synthetic (verbal or secondary) compounds and root (non-verbal or primary) compounds. According to Selkirk (1982), verbal compounds are "endocentric adjectives or noun compounds whose head adjective or noun (respectively) is morphologically complex, having been derived from a verb, and whose non-head constituent is interpreted as an argument of the head adjective or noun" (p. 23). Arguments are defined as "a noun phrase bearing a specific grammatical or semantic relation to a verb and whose overt or implied presence is required for well-formedness in structures containing that verb" (Trask, 1993, p. 20). They can be identified in two ways: (1) in terms of syntactic roles with respect to the verb, such as Subject and Object, and (2) in terms of semantic roles in relation to the verb, such as AGENT (entity that instigates an action) and PATIENT (entity that undergoes an action).

Argument structure is relevant to the classification of compounds as only in synthetic (or verbal) compounds is the first component regarded as an argument of the head, that is, as a core component of the meaning. For example, *time-saver*, *house-cleaning*, and *water-repellent* are all synthetic compounds, since the first component of all these compounds is an argument of the second.

Compared to synthetic compounds, root (non-verbal or primary) compounds are those in which "the non-heads add a locative, manner, or temporal specification to the head, but would not be said to bear a thematic relation to, or satisfy the argument structure of, the head" (Selkirk, 1982, p. 24). For instance, in *party drinker* the first noun expresses location, in *hardworking*, *hard* denotes manner, and in *long-suffering* the adjective *long* expresses time. In all these cases, the participants are circumstantials, as they add peripheral information and are optional for characterising the meaning of the head noun.

Quirk et al. (1985, pp. 1570–1578) provide a comprehensive classification resulting in four main patterns for noun compounds and three patterns for adjective compounds, which are further subdivided into more specific sets. The main classes are shown in Table 2.

Table 2. Syntactic classification of English compounds (Quirk et al., 1985)

Compound word-class	Compound internal structure	Examples
Noun compounds	Subject + Verb	*sunrise*
	Verb + Object	*taxi driver*
	Verb + Adverbial	*swimming-pool*
	Verbless	*armchair*
Adjective compounds	Verb + Object	*eye-catching*
	Verb + Adverbial	*heart-felt*
	Verbless	*footsore*

As can be seen in Table 2, Quirk et al. (1985) classify noun and adjective compounds into different sets, depending on the syntactic relationship held between their components. For example, while in the noun *taxi driver* and the adjective *eye-catching* the first constituent can be analysed as the object of the second one, in the noun *swimming-pool* and the adjective *heart-felt* the function of the first element is adverbial, as *swimming-pool* is interpreted as "swimming in the pool" and *heart-felt* as "felt in the heart".

Another relevant classification is the one proposed by Bisetto and Scalise (2005). The authors distinguish between attributive compounds, whose components hold a modifier–head relation, like *snail mail, windmill, school book*, and subordinative compounds, where one component is the argument (subject or object) of the other, that is, they are synthetic in Selkirk's (1982) terminology. Examples of this group are *mystery writer, food shopping, home invasion*, and *cut throat*. Finally, the authors distinguish coordinative compounds, where both components enjoy equal weight, such as *producer-director, doctor-patient*, and *blue-green*.

As can be inferred from the previous classifications, in spite of their differences, they show consensus on a basic distinction between compounds where the first noun stands in an argument-like/object relationship to the second noun, and those where the first noun holds a more peripheral type of association, that is, adverbial-like. This basic twofold distinction is again confirmed by Bauer et al.'s (2013) classification of compounds into argumental and non-argumental.

The syntactic classification intersects with the semantic one in that the same types of syntactic relations can be found in both endocentric and exocentric compounds, as illustrated in Table 3.

Table 3. Syntactic classification across semantic types

Endocentric	Exocentric	Syntactic class
truckdriver	*scarecrow*	Subordinative
bookcase	*loudmouth*	Attributive
producer-director	*parent-child*	Coordinative

In addition to their syntax-like internal structure, compounds also share the possibility of recursiveness with syntactic units. This implies that a compound formed with two bases, like *paper towel*, can be combined with another base, such as *paper towel dispenser*. The resulting compound can then be joined with yet another base, leading to a chain of compounds, like *paper towel dispenser factory*, and so on. Ultimately, this recursive process allows for the creation of highly complex compounds, such as *paper towel dispenser factory building committee report*.

The focus in the present study will be some specific types of Quirk et al.'s (1985) group of noun compounds. Within their set of Verb + Object compounds, they include one group that they analyse as Object + Verbal Noun in *-ing* (e.g., *fault-finding*). In addition, within their set of Verb + Adverbial compounds, they suggest different groups. One of them is formed by compounds with the structure Adverbial + Verbal Noun in *-ing*, like *sun-bathing*, *daydreaming*, and *handwriting*. Since the aim is to explore the use of English compounds to designate adventure tourism activities, it does not seem counterintuitive to restrict the analysis to these two specific groups containing a noun in the *-ing* form. The addition of the suffix *-ing* to any English verb (and in certain cases, nouns) yields nominal forms referring to the activity associated with the verb's meaning. For example, *swimming* is defined as "the activity or sport of swimming" (Cambridge Dictionary, n.d.), while *bicycling* means "the activity of riding a bicycle" (Cambridge Dictionary, n.d.). Occasionally, the base may even not be an English verb, as exemplified by *sea kayaking*, which is not derived from the verb **sea kayak*. In any case, by restricting the analysis to these cases, it was reasonable to anticipate the retrieval of a substantial number of instances.

3. Methodology: Data Collection and Analytical Procedure

The data used for the current research were retrieved from ADVENCOR, a specialised online bilingual corpus containing similar texts in Spanish and English, which was compiled for the project in which this chapter is framed (cf. Durán-Muñoz & Jiménez-Navarro, 2021). The English subcorpus has 1,064,664 words, and the Spanish one has 1,118,903 words. Both subcorpora are balanced and are representative of adventure tourism. They were compiled semi-automatically using Sketch Engine (Kilgarriff et al., 2004), but the process was supervised thoroughly according to the authors (Durán-Muñoz & Jiménez-Navarro, 2021).

The search for the data in this study was conducted using the Sketch Engine *N-gram* function in order to produce frequency lists of uninterrupted multi-word expressions. By asking for word + word ending in *-ing*, 2,640 combinations were retrieved. False positives were deleted and duplicates (same word, different orthographic realisations) were removed. The head nouns of the sequences obtained were then searched using the *Word Sketch* function to avoid missing any of the sequences of word + *-ing* word. This resulted in a total number of 400 compounds, which were then classified in terms of the activity denoted by the head noun, that is, the second element. In this regard, although most of the compounds designated an activity subsumed under adventure tourism, sometimes other facets were expressed. Thus, some of the compounds referred to a preceding stage (e.g., *online booking, parachute training, tour planning, welcome meeting, team building, safety briefing*), while others focused on one specific action or episode that a tourism activity involves (e.g., *map reading, pack carrying, parachute opening*), but not to the activity per se. Figure 1 shows the different stages in the adventure tourism semantic frame for which compounds were found.

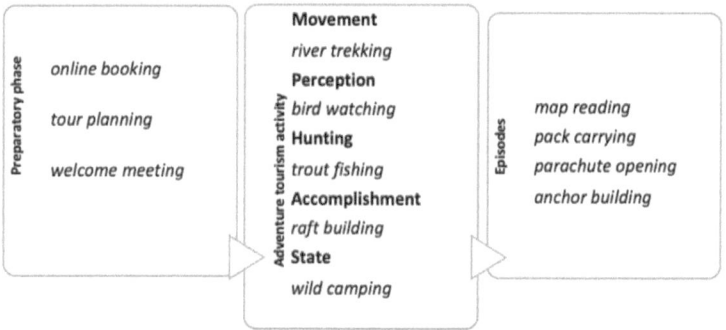

Figure 1. Adventure tourism frame in the compound data

For the sake of simplicity, compounds that did not denote an adventure tourism activity were excluded from the analysis, that is, only those in the middle rectangle of Figure 1 were explored. This left a final total amount of 322 compounds, which were subjected to the analysis shown in Section 4.

Table 4 displays a sample of the analysis conducted with the compound data. The compounds were listed (column 1) and coded by looking at the type of entity denoted by the first word, as shown in column 2. In the third column, the internal structure of the compounds in terms of the word class of their components was provided. Column 4 includes the head nouns (a total of 61 different nouns), which are followed, in column 5, by the head type number, that is, the number of compounds with the same head and different first constituents in the data.[1] For example, there are three types with *backpacking* in head position (*lightweight backpacking, overnight backpacking,* and *ultralight backpacking*), while there are two types with *ballooning* (*cultural ballooning* and *hot air ballooning*). Finally, column 6 provides the internal semantic relation held between the first word and the head noun, that is, the semantic role of the first noun or adjective (and occasionally, verb) in the activity designated by the head noun.

Table 4. Compound analysis sample

	Semantic class of the 1st word	Structure	Head noun activity	Type No.	Semantic role of the 1st word
lightweight backpacking	property	NN	backpacking	3	Property
overnight backpacking	property_time	AN			Time
ultralight backpacking	property	AN			Property
cultural ballooning	property_purpose	AN	ballooning	2	Property
hot air ballooning	nature_substance_air	[AN]N			Instrument
bush bashing	plant_bush	NN	bashing	3	Path
dune bashing	nature_dune	NN			Path
ridge bashing	nature_ridge	NN			Path
mountain bicycling	nature_mountain	NN	bicycling	1	Path

(*continued*)

---

1 When measuring morphological productivity using corpus data, the term "type" is used to refer to all the different derived words which result from the attachment of a suffix to different bases (Plag, 2003, p. 50). The term "token" refers to the number of instantiations of each type. In this paper, compounds with the same head and a different first constituent are regarded as types. These types are to be distinguished from the number of occurrences of each type, that is, the tokens, which are not the concern in this study.

Table 4. Continued

	Semantic class of the 1st word	Structure	Head noun activity	Type No.	Semantic role of the 1st word
dirt biking	nature_dirt	NN	biking	4	PATH
fat biking	property	AN			PROPERTY
mountain biking	nature_mountain	NN			PATH
quad biking	artefact_vehicle_quad	NN			VEHICLE_WITH_ENGINE
body boarding	nature-body	NN	boarding	11	INSTRUMENT
Cerro Negro sand boarding	nature_name_volcano	N[NN]			PATH
fly boarding	action_fly	VN			ACTION
mountain boarding	nature_mountain	NN			PATH
Pacaya volcano boarding	nature_vname_volcano	[NN]N			PATH
paddle boarding	artefact_instrument_paddle	NN			INSTRUMENT
stand-up paddleboarding (SUP boarding)	action	V[NN]			ACTION
wake boarding	nature_wake	NN			PATH
river boarding	nature_river	NN			PATH
sand boarding	nature_sand	NN			PATH
snow boarding	nature_snow	NN			PATH

4. Analysis and Discussion

In this section, an in-depth analysis of the compounds is carried out from both syntactic and semantic perspectives. The syntactic analysis (Section 4.1.) starts with a discussion of some controversial cases in Section 4.1.1., followed by their categorisation according to their internal structural composition in Section 4.1.2., specifically focusing on the grammatical category of their constituent elements. A semantic exploration follows in Section 4.2., in which the compound data are subjected to scrutiny along various dimensions, including the head noun denotation (Section 4.2.1.), the semantic class of the left-hand side word (Section 4.2.2.), and the semantic relation between the two component words in terms of thematic roles (Section 4.2.3.).

4.1. Syntactic analysis

4.1.1. Compound or phrase

As mentioned in Section 2.1., compounds are not always a straightforward category, as there are cases where the distinction from other types of units is not crystal clear. In the corpus data, some of these fuzzy cases were encountered.

Firstly, there are examples where the first word is an evaluative adjective, like *good walking, easy climbing, exciting rafting*, and *challenging trekking*. In these cases, the adjective does not classify the activity into different categories, as it does in compounds with non-evaluative adjectives (e.g., *guided fishing, technical climbing*, and *free flying*), but the adjective provides subjective evaluation of the activity instead. However, an evaluative adjective can also specify the tourism activity objectively, such as in *brisk walking, strenuous walking, short trekking*, or *high diving*, which does not help in making the identification of true compounds clearer. A number of these cases were included among the compound data, as long as they are recognised as lexical units in web sources.[2]

A second group of unclear sequences is formed by examples with a proper name in the left-hand side position, in particular, names of geographical entities. The analysis of these cases as compounds is not without problems, and some scholars might not agree on considering combinations like *Everest trekking* as a morphological unit, that is, a compound. Nevertheless, the issue of distinguishing two types of Noun–Noun (or Adjective–Noun) constructions is still unresolved, and it is not the aim of this work to make an attempt to solve the puzzle by applying the traditional criteria, which has been shown to fail in the morphological research on the area (Bauer, 1998). Aware as we are of the underlying theoretical problem, for the present purposes some of these sequences were regarded as compounds. For example, *Everest trekking, Barron rafting*, and *Cerro Negro sand boarding* designate popular adventure tourism experiences that are thoroughly described in specialised sources. The natural surroundings, climate, and local culture of specific locations where some adventure tourism activities are carried out can significantly influence their identity and appeal, making them unique experiences worthy of a name, that is, a compound noun. By contrast, proper names, referring to adventure tourism companies (e.g., *Alpine Rafting, Northeast Mountaineering, Green Mountain Skydiving, Pokhara paragliding*, and the like) were excluded.

---

2   See, for example, *brisk walking*: https://www.collinsdictionary.com/es/diccionario/ingles/brisk-walking [Last accessed: 28/09/2023].

A third controversial case is the set of sequences including a numeral or a quantifier and a time-denoting first noun (e.g., *a one-day canyoning, two nights camping*). Certain cases among these lend themselves to analysis as compounds, indicating an activity specified for duration (e.g., *a half or full-day kayaking, ½ day rafting*). However, the analysis of these cases as compounds is dubious, and the majority of the sequences are partitive phrases expressing the amount of time devoted to the adventure activity (e.g., *four days kayaking*, meaning "four days of kayaking"). Therefore, this set was excluded from the analysis.

Finally, as mentioned in Section 2.1.2., one property of compounds that makes them similar to syntactic units is recursiveness, that is, the possibility of including one compound into a larger one, which results in three or four (or even larger) word units. Some examples of recursiveness were found in the data. One of these is *recreational marine life showcasing*, where *marine life* is a compound which is made part of a new larger compound formation, rendering *marine life showcasing*, a compound with another compound noun in its head position. This is again used as the input of an even larger unit, resulting in *recreational marine life showcasing*. Serial compounding highlights the blurred distinction between morphological and syntactic units, as the longer a formation becomes the more challenging it is to classify it as a lexical unit (i.e., a compound). Further examples like this one are *wind tunnel indoor skydiving, vertical wind tunnel skydiving, outrigger-canoe surfing, helicopter flightseeing, bungee trampoline riding,* and *Annapurna base camp trekking*.

4.1.2. Syntactic category of the first word

Noun–Noun compounds (e.g., *mountain skiing*) were found to be more frequent than Adjective–Noun compounds (e.g., *alpine skiing*), representing 63.35 % and 25.77 % of the compounds, respectively.

When the first component is an adjective, the function is that of a modifier, so these compounds are regarded as non-argumental, attributive, or root according to the syntactic classification provided in Section 2.1. However, there are cases where the semantic role of the adjective is similar to that of a noun. An example is *aerial trekking*, which designates a subtype of trekking that takes place in the air, where the adjective fulfils the role of PLACE.

In addition to the more frequent Noun–Noun and Adjective–Noun patterns, some of the sequences are examples of recursiveness, as mentioned in Section 4.1.1. Among these cases, different combinations are found, such as [NN]N (e.g., *dog sled racing, air gun shooting, motor car racing, Pacaya volcano boarding*), [AN]N (e.g., *African wildlife viewing, deep sea fishing, dirst track racing*), and

A[NN] (e.g., *bareback horse racing*). Complex compounds like these make 9.62 % of the data.

Finally, there are a few compounds where the first component is verbal, such as *lead climbing*: "a technique in rock climbing where the *lead climber* ascends the climbing route clipping their rope to climbing protection as they progress, while their *second* (or belayer) remains at the base of the route belaying the rope to protect the *lead climber* in the event that they fall",[3] *hang gliding*: "the activity of flying through the air by hanging from a very small aircraft without an engine, consisting of a frame covered in cloth" (Cambridge Dictionary, n.d.), or *stand up paddle boarding*: "a subclass of paddleboarding, a broader concept that also includes the use of arms while kneeling, lying, or standing on a narrow and long paddleboard to move around in the water".[4] This pattern is notoriously less frequent than Noun–Noun or Adjective–Noun compounds, representing scarcely 1.26 % of the data. The different patterns encountered are shown in Table 5.

Table 5. Internal structure of the compounds in the data

Compound structure	Example
NN	*cave tubing*
AN	*Nordic walking*
[NN]N	*dog sled racing*
N[NN]	*helicopter flightseeing*
[AN]N	*oval track racing*
A[NN]	*bareback horseracing*
[NN][AN]	*wind tunnel indoor skydiving*
[N[NN]]N	*Everest base camp trekking*
[A[NN]]N	*vertical wind tunnel skydiving*
VN	*lead climbing*
V[NN]	*stand-up paddle boarding*
A[[AN]N]	*recreational marine life showcasing*
[AAN]N	*great white shark diving*

---

3   https://en.wikipedia.org/wiki/Lead_climbing [Last accessed: 28/09/2023].
4   https://www.surfertoday.com/surfing/what-is-stand-up-paddleboarding-sup [Last accessed: 28/09/2023].

4.2. Semantic analysis

4.2.1. Head noun denotation

Compounds designating specific tourism activities were classified into different sets (cf. Figure 1). The first group (which is by far the most frequent one) includes all nouns denoting some kind of movement (83.22 %), either on the ground (e.g., *gorge walking, free running, backcountry skiing, mountain trekking, alpine hiking, track racing*), on or in water (e.g., *blackwater rafting scuba diving, river kayaking*), on ice (e.g., *ice climbing*), or by air (e.g., *aircraft flying, tandem paragliding*). Some of these head nouns denoting movement are metonymic in that the noun that designates an instrument, such as a vehicle (and, more marginally, a natural entity) (e.g., *canoe, raft, boat, parachute, ski, balloon, bicycle, bike, backpack, rope, kayak, paddle, sled, canyon*), stands for the action in which the instrument is involved (e.g., *open canoeing, paddle boarding, jet boating, river rafting, air ballooning, top roping, canopy kayaking, dog sledding*).

The remaining semantic sets are: (1) hunting (6.52 %), including nouns that designate the activity of chasing or fruit collecting (e.g., *game hunting, trout fishing, pigeon shooting, mushroom hunting, berry picking*); (2) stative activities (5.60 %), with nouns derived from state verbs, like *wild camping* or *country living*; (3) perception (4.03 %), which subsumes nouns involving a stimulus and a perceptual activity (e.g., *wildlife spotting, wildlife viewing, bird watching, sightseeing, star gazing*); and, finally, (4) some deverbal nouns that denote other resultative actions (different from hunting), that is, accomplishments (0.63 %), and are the head of compounds such as *raft building* or *picture taking*.

The same head nouns are used in different types, resulting in different compound words which designate distinct adventure tourism activities. For instance, in addition to *valley trekking*, we find *elephant trekking, aerial trekking*, or *Everest trekking*, among other subtypes of trekking. The top-10 head nouns in terms of the number of compound types of which they are part are displayed in Figure 2.

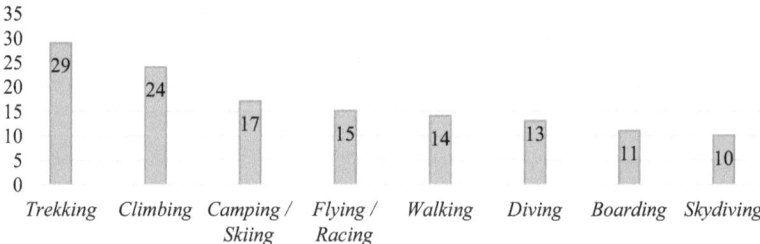

Figure 2. Top-10 head nouns (type number)

The use of compounds in the adventure tourism lexicon 243

The compound types with *trekking*, *climbing*, and *skiing*, that is, all the compounds containing each of the previous top three head nouns and a different first constituent, are provided in Table 6.

Table 6. Compound types with the top three head nouns

	Trekking	Climbing	Skiing
1	adventure trekking	adventure climbing	alpine skiing
2	aerial trekking	beginner climbing	backcountry skiing
3	altitude trekking	face climbing	barefoot skiing
4	Annapurna base camp trekking	free climbing	cross-country skiing/ x-county skiing
5	Annapurna trekking	guided climbing	downhill skiing
6	Annapurna region trekking	ice climbing	extreme skiing
7	circuit trekking	indoor climbing	freeride skiing
8	coastal trekking	lead climbing	heli skiing
9	Dolpo trekking	low-level climbing	jet skiing
10	elephant trekking	mixed climbing	kite skiing
11	Everest trekking	mountain climbing	mountain skiing
12	Eveerst base camp trekking	multipitch climbing	ordinary skiing
13	ferrata trekking	outdoor climbing	off-piste skiing
14	Hampta Pass trekking	peak climbing	sand skiing
15	Himalayan trekking	rated climbing	snow skiing
16	hut trekking	rock climbing	summer skiing
17	Larke Pass trekking	rope climbing	water skiing
18	Makalu region trekking	snow climbing	
19	Manaslu trekking	sport climbing	
20	Mount Rinjiani trekking	technical climbing	
21	mountain trekking	traditional climbing/trad climbing	
22	Mustang trekking	tree climbing	
23	Nepal trekking	vertical climbing	
24	pony trekking	via ferrata climbing	
25	professional trekking	wall climbing	
26	river region trekking		
27	short trekking		
28	snow trekking		
29	tandem trekking		
30	valley trekking		

As can be seen in Table 6, some head nouns are particularly significant in the formation of compound types, with *trekking* occupying the top position, being found in 9 % (29 types) of the total data, followed very closely by *climbing*, which is a component of 7,45 % (24 types) of the compounds. *Camping* and *skiing* are the next most frequent head nouns, and each of them is found in 5.27 % (17 types) of the corpus data.

4.2.2. Meaning of the first noun or adjective

As mentioned in Section 3, words that occupy the left-hand side of the compound were categorised into different sets in terms of the entity type denoted. The vast majority of words in the first slot position are nouns denoting first-order entities, to put it differently, entities that are relatively constant in terms of their perceptual properties and are publicly observable. According to Lyons (1977), animals, people, and discrete physical objects correspond to this description. In this work, nouns designating natural entities occupy the top position, which include elements of nature like *mountain* (e.g., *mountain biking*), *valley* (e.g., *valley trekking*), *river* (e.g., *river rafting*), *gorge* (e.g., *gorge walking*), but also substances like *water* (e.g., *water surfing*), *sand* (e.g., *sand skiing*), *snow* (e.g., *snow trekking*), or *ice* (e.g., *ice skating*), as well as animals (e.g., *horseback riding*), people (e.g., *family* rafting), and plants (e.g., *bush bashing*) or earth products (e.g., *mushroom picking*). Some nouns referring to natural entities are proper names, such as mountains (e.g., *Manaslu trekking*) or rivers (e.g., *Barron rafting*). As mentioned in Section 4.1.1., these sequences designate specific subtypes of the general activity denoted by the head and are conceptualised as such by practitioners of adventure tourism, which motivates their analysis as compound-like sequences. Thus, certain regions or mountains are renowned places where trekking is practised, so *Annapurna trekking, Dolpo trekking*, or *Everest trekking* are compounds that name these different trekking experiences.

The set of compound nouns with nouns denoting natural entities in the first slot is followed in frequency by compounds with an adjectival first component designating a property (e.g., *extreme canyoning*), a semantic category different from first-order entities (Hengeveld & Mackenzie, 2008). Adjectives can also refer to natural entities (e.g., *alpine skiing*) or to places (e.g., *base jumping, cross-country driving, indoor climbing, off-piste skiing*), which constitute a different type of entity in Lyons's (1977) typology. Compounds with a first noun denoting an artefact, that is, a second type of first-order entity, follow those with natural entity first words in number. This set includes not only any man-created entity (e.g., *bridge crossing, pond swooping*) but also instruments (e.g., *kite surfing*) and vehicles (e.g., *helicopter flightseeing, tandem skydiving*). The least relevant sets in terms of type frequency are those where the first noun designates a second-order entity, that is, an action (e.g., *stand-up paddle boarding*) or a temporal entity (e.g., *overnight camping*). The different semantic classes of the first component word are shown in Table 7.

Table 7. Frequency of N_1 semantic class

Word 1 semantic class		Example	Frequency
First-order entity	Nature	*waterfall rappelling, water skiing, gorge scrambling*	116
	Artefact	*pond swooping, kite surfing, oval track racing*	55
	Animal	*game hunting, horseback riding, whale watching*	21
	Plant	*mushroom hunting, berry picking, bush bashing*	5
	Person	*family rafting, small group walking, family camping*	3
Property		*free running, recreational flying, trad climbing*	84
Place		*back-country camping, off-piste skiing, indoor climbing*	28
Time		*winter tubing, overnight camping, summer skiing*	4
Second-order entity		*stand-up paddle boarding, lead climbing, fly boarding*	6

4.2.3. Semantic relation of $N_1/A_1/V_1$ in activity denoted by N_2

In Section 2.1., a distinction was drawn between argumental and non-argumental compounds. In the specific area of adventure tourism, a distinction can, a priori, be drawn between these two types, depending on the different roles of the participants involved in the verbs from which the head noun has been derived. It must be remembered that argumental (synthetic or verbal) compounds display actant participants (i.e., arguments or elements required for the meaning of the verb to be complete) in their first slot, whereas non-argumental (root or non-verbal) compounds exhibit circumstantial participants (i.e., participants that are more peripheral to the meaning of the verb) as their first component. Since compounds with an adjective in the first position do not show argument-like relations in general, these compounds can be analysed as compounds where the head noun is specified by a modifier denoting a property (manner, purpose, height, and the like).

As regards Noun–Noun compounds, the first noun can be analysed as a participant in the activity denoted by the head noun. The most frequent core participants in the specialised domain of adventure tourism are the people involved (Agent), the Location where the activity takes place, and the Instrument required in order to carry out the activity, as they are the key elements of the conceptual representation of any adventure activity (Figure 3) (Durán-Muñoz, 2016).[5]

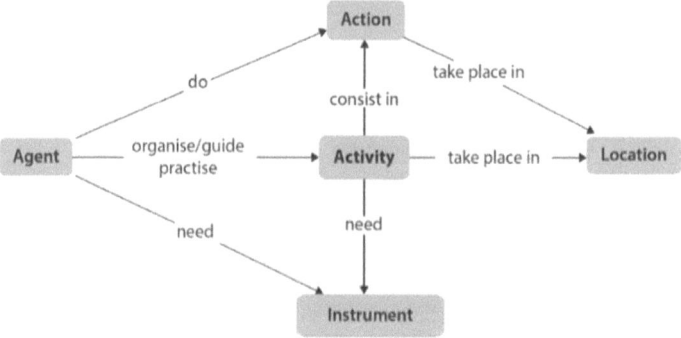

Figure 3. Prototypical conceptual representation of an adventure activity (Durán-Muñoz, 2016)

However, these basic thematic relations (or semantic roles) have been refined in the field of adventure tourism in order to account more accurately for the more specific meaning relations that can be expressed (Durán-Muñoz & L'Homme, 2020, p. 46). Thus, the semantic role ("conceptual category" in Durán-Muñoz & L'Homme's [2020] terminology) Agent is subdivided into Tourist and Responsible. The category Location subsumes Path, Place, Source, Direction, and Destination. Finally, the semantic role Instrument is subdivided into

---

5   Durán Muñoz and L'Homme (2020) propose a terminological description of motion verbs extracted from the Advencor corpus. Their descriptive model aims to account for the different relations between verbs and their participants in a systematic way. This model is based on a lexico-semantic analysis proposed by L'Homme (2012, 2018) and implements some principles of Frame Semantics (Fillmore, 1976, 1982; Fillmore & Baker, 2010) in that all participants (arguments and circumstantials) are labelled in such a way that their "association in experience" (Croft & Cruse, 2004, p. 7) is made explicit.

Instrument, Vehicle_with_Engine, Vehicle_without_Engine, Safety_Instrument, and Clothing. This finer-grained distinction allows differentiating between *scuba diving, scooter riding, raft building, shark cage diving,* and *wingsuit*, where the first nouns denote an instrument, a vehicle with engine, a vehicle without engine, safety equipment, and an article of clothing, respectively.[6]

It is expected that some of these roles are expressed more productively than others. For example, since most tourism activities involve movement through water, air, or on the ground, it is not surprising that Path or Place are among the most frequent roles, representing 25.80 % and 16.14 % of the data, respectively. It should be noted, however, that in actual practice the distinction between these specific roles might not be easy to make in some cases. For instance, in *snow boarding*, *snow* can be analysed as the Path or the Place where the activity is carried out. In any case, subtle distinctions make it possible to determine which semantic relations are more likely to make their way into compound formations. The Location semantic role, for its part, is mainly illustrated by the Path (e.g., *via ferrata climbing, road cycling, bridge crossing, ocean kayaking, track racing, water rafting, waterfall rapelling*) and Place (e.g., *city skydiving, Alpine skiing, beach camping*) subroles, with Source (e.g., *cliff jumping*) and Destination (e.g., *peak climbing, Everest base camp trekking*) being scarcely represented in the corpus data (0.62 % and 2.20 %, respectively). In the case of the Instrument role, 9.62 % of the corpus data illustrate the Instrument subrole (e.g., *scuba diving, bungee jumping, kite surfing, cable gliding, paddle boarding*), and the Vehicle_with_Engine (e.g., *boat fishing*) or Vehicle_without_Engine (e.g., *dog sled racing*) subroles constitute 7.80 %, while the Safety_Instrument (e.g., *canopy kayaking*) and Clothing (e.g., *wingsuit flying*) subroles are not particularly relevant in compound formation (0.6 % each).

In addition, this analysis has revealed the requirement of further semantic roles to account for compound formation. Thus, 24.22 % of the corpus data are compounds where the first word specifies the head noun by means of a Property, like the Manner (e.g., *barefoot skiing, brisk walking, free falling, tandem parasailing*), the Purpose (e.g., *recreational fishing, rescue swimming, adventure trekking, competitive racing*), or the Level_of_Difficulty (e.g., *extreme skiing, basic paddling*) of the activity. A Stimulus role is performed by the first word of compounds that refer to activities of perception, such as *whale watching, star viewing*, or *sunset viewing*, but also in other cases, like *shark cage*

---

6   The different roles subsumed under each of the basic thematic relations Agent, Location, and Instrument are defined in Durán-Muñoz and L'Homme (2020, p. 46).

diving and *great white shark diving*, amounting to 4 % of the data. The head of some compounds can also be specified by a noun designating a TOURIST (e.g., *family rafting*), representing a scarce 1.22 %, a PATIENT (e.g., *duck hunting*), with 1.86 % of the data, a RESULT (e.g., *raft building*), with 2.20 %, an ACTION (e.g., *hanggliding*), with 1.86 %, or a TIME unit (e.g., *summer skiing*), with another 1.86 %. These findings show that semantic roles which are usually optional, like MANNER, PURPOSE, or TIME, might be core in compounding.

In terms of the distinction between synthetic (or argumental) and primary (or non-argumental) compounds, this lack of one-to-one correspondence between specific semantic roles and their realisation as actants or circumstantials means that compounds displaying similar thematic relations should be classified as synthetic in some cases and as root compounds in others, despite being the same type of unit in actual practice. In other words, the element is always present regardless of whether it is an actant or a circumstantial, unlike in the case of syntactic units, where an element classified as circumstantial is optional.

Regarding the relation between semantic roles and the entity type denoted by the first noun in the tourism activity, there is not one-to-one correspondence. For instance, a first noun designating a substance, such as ice, can be the PATH through which the activity takes place (e.g., *ice skating*) or a PLACE in a different compound (e.g., *ice fishing*). A similar example is the semantic role of *rock* in *rock climbing*, where *rock* is the PATH, as compared with *rock skydiving*, where *rock* is the SOURCE. Conversely, the same thematic role can be realised by different types of entities. This is illustrated with the semantic role PATH in the Appendix, which shows its realisations by means of a natural entity (e.g., *dune bashing, mountain biking, river crossing, tree climbing, canyon hiking, gorge paddling, snow climbing*), an artefact (e.g., *road cycling, circuit trekking, track racing*), or a property (e.g., *cross-country skydiving, downhill skiing*).

4.2.4. Semantic transparency

In Section 2, semantic unpredictability was alluded to as a criterion that characterises English compounds. However, the analysis of the corpus data in this study has revealed that opacity is generally not a diagnostic feature for compoundhood in the case of the adventure tourism domain. Most of the compounds analysed are transparent, since the head noun refers to an activity where the first noun usually plays one of the roles participating in the tourism event so that the meaning can be parsed with ease in the majority of cases.

Nevertheless, opaque examples can be found occasionally. Thus, *white-water rafting* does not lend itself to a straightforward interpretation as "the activity of being moved quickly in a raft (= small boat filled with air) along rivers where the current is very strong" (Cambridge Dictionary, n.d.). Similarly, "the activity of riding in an inner tube (= large rubber tube filled with air, like those inside a tire) along a fast-flowing river that runs under the ground through caves" is referred to as *blackwater rafting* (Longman Dictionary of Contemporary English, n.d.), so the meaning might not be accessed straightforwardly either. Further examples are *fat biking*, where the adjective's property is not attributed to the head noun activity but to the bicycle's tyres, *Nordic walking*, referring to a Finnish-origin walking technique that can be performed anywhere, or *lead climbing*, difficult to understand unless we know it involves a leader, that is, the first person climbing.

In some cases, opacity can be explained by the fact that a metonymic process is at work and must be activated to trigger the compound's interpretation, as mentioned in Section 4.2.1. For instance, in a few examples, such as in *jet boating, open canoeing, fat biking, dog sledding, hot air ballooning, zip lining, inner tubing,* or *top roping*, the tourism activity is accessed via the vehicle or the instrument participating in the activity, since there is no verbal base from which the head noun has been derived by -*ing* suffixation. In other cases, world knowledge facilitates the correct interpretation. This is the case of *(shark) cage diving*, "an activity in which people are taken underwater in a cage so that they can see sharks (= very large fish with sharp teeth) swimming near them" (Cambridge Dictionary, n.d.), where the STIMULUS role of the shark in the diving activity is retrieved from world experience.

Ambiguity is not very common but can be found in some examples, such as *canopy kayaking*, which can refer to the activity of moving in a kayak with a cover fixed over it for shelter or, more unexpectedly, to the place or path for the activity. This second meaning is displayed in the following contextualised example extracted from ADVENCOR: "*Canopy kayaking* in the flooded forest. How is it possible to kayak in the canopy?" In other cases, our world knowledge facilitates access to the meaning, as in *windsurfing*, "a sport in which you sail across water by standing on a board and holding onto a large sail" (Cambridge Dictionary, n.d.).

Opacity increases when the first word does not express a semantic role. For instance, one can figure out that *mixed climbing* involves some kind of mix, but the actual meaning, which is a combination of ice climbing and rock climbing, is difficult to be accessed unless we have some expertise in climbing. Except for this and similar cases, repeated instantiations of the different semantic patterns result in their entrenchment in the mental lexicon, which permits semantic predictability and emergence of new cases.

5. Conclusions

The research presented in this chapter constitutes a multidisciplinary study where linguistic theory and description have been applied to research on terminology. Specifically, the analysis of the corpus data shows the relevance of the process of compounding in the specific lexicon of adventure tourism. Since motion verbs are crucial in this area, one set of compounds was explored, those with a deverbal head noun ending in *-ing*.

The productivity of compounds evinces the rising popularity of adventure tourism. As pointed out by Durán-Muñoz and L'Homme (2020), people's interest in sport, nature, and sustainability is on the rise, so traditional vacations are being replaced by more active ones. The lexicon used to designate adventure tourism activities is the reflection of this new real-world trend.

The analysis of real data casts light on the patterns that are more frequent and are thus entrenched in the minds of language users, making it easy the interpretation and creation of new instantiations. Syntactically, Noun–Noun compounds have been shown to be most productive, amounting to 63.35 % of the total number, followed by Adjective–Noun compounds, representing 25.77 % of the data, and Verb–Noun compounds, which are very scarce, with 1.26 % of the examples. The remaining 9.62 % of the data are complex compounds illustrating recursiveness.

Semantically, most productive compound types in terms of head noun frequency are those that designate movement, which comes as no surprise, bearing in mind that movement is almost inherently associated to adventure tourism (e.g., *wall climbing, recreational flying, tandem paragliding, alpine skiing, bungee jumping, hill walking, coastal trekking, outdoor swimming, backcountry running*). This semantic set includes 83.22 % of the data, while the remaining sets are only marginally represented, with 6.52 % compounds denoting hunting or fruit harvesting activities (e.g., *pigeon hooting, ice fishing, game hunting, mushroom picking*), 5.60 % compounds expressing states (e.g., *night camping, winter camping, country living*), 4.03 % compounds designating perception (e.g., *whale watching, star gazing, sightseeing, storm chasing*), and an infrequent 0.63 % of compounds that express accomplishments (e.g., *picture taking, raft building*). As regards semantic roles that connect the two constituents forming the compound, PATH, PROPERTY, and PLACE are by far the most relevant relations, representing 25.80 %, 24.22 %, and 16.14 % of the data, respectively.[7]

---

[7] PATH is also a relevant semantic role in a different type of phraseological units, more specifically, in prepositional phrase collocations of motion verbs, where the prepositional phrase commonly expresses this role (cf. Jiménez-Navarro, this volume).

The results of the present study provide useful material in the compilation of specific terminology in the domain of adventure tourism. As a matter of fact, the seed for the research conducted in this chapter is a recent project aiming at designing and developing an online bilingual dictionary of adventure tourism vocabulary known as *DicoAdventure*.[8] This dictionary is intended as a flexible and encompassing resource to allow the acquisition of specialised lexicon intuitively by using semantic frames. The 322 compound words that have been explored in this chapter are a modest contribution for the awareness of the specialised terminology of adventure tourism.

The notorious use of compounding in this domain can be seen as motivated by the search for economy of expression characterising the language of specialised terminology and is, thus, expected to be found in further domains, such as computing or law. Future lines of research can pursue the investigation of compounding in these domains, where it might be proven to be equally relevant. In addition, a more thorough understanding of the adventure tourism segment might be achieved by expanding the study to further morphological processes or linguistic areas.

Finally, the results of the data analysis have theoretical implications concerning common knowledge of morphological tradition of compounding types. Adverbial-like relations have been shown to be actants in adventure tourism verbs, and these relations are passed on to compound nouns, where the left-hand side noun plays an argument-like role (i.e., actant) in the activity designated by the head noun. The linguistic literature on compounding has drawn a distinction between synthetic (verbal, secondary, or argumental) and root (primary, non-argumental) compounds, based on the presence of an argument in the former or an adverbial relation in the latter. However, the relevance of adverbial-like relations playing an argument role in motion verbs and the corresponding derived compound nouns renders the traditional distinction vacuous.

References

Adams, V. (1973). *An introduction to modern English word-formation.* Longman.

Adams, V. (2001). *Complex words in English.* Pearson Education.

Bauer, L. (1983). *English word formation.* Cambridge University Press.

Bauer, L. (1998). When is a sequence of two nouns a compound in English? *English Language and Linguistics, 2*(1), 65–86.

---

8   http://olst.ling.umontreal.ca/dicoadventure/ [Last accessed: 28/09/2023].

Bauer L. (2001). Compounding. In M. Haspelmath, E. König, W. Österreicher, & W. Raible (Eds.), *Language typology and language universals, vol. I* (pp. 695–707). Mouton de Gruyter.

Bauer, L., Lieber, R., & Plag, I. (2013). *The Oxford reference guide to English morphology*. Oxford University Press.

Bisetto, A., & Scalise, S. (2005). The classification of compounds. *Lingue e Linguaggio*, 4(2), 319–332.

Cambridge Dictionary. (n.d.). bicycling. In *Cambridge Dictionary*. Retrieved August 24, 2023, from https://dictionary.cambridge.org/dictionary/english/bicycling

Cambridge Dictionary. (n.d.). cage diving. In *Cambridge Dictionary*. Retrieved September 10, 2023, from https://dictionary.cambridge.org/dictionary/english/windsurfing

Cambridge Dictionary. (n.d.). hang gliding. In *Cambridge Dictionary*. Retrieved August 1, 2023, from https://dictionary.cambridge.org/es/diccionario/ingles/hang-gliding.

Cambridge Dictionary. (n.d.). swimming. In *Cambridge Dictionary*. Retrieved August 24, 2023, from https://dictionary.cambridge.org/dictionary/english/swimming

Cambridge Dictionary. (n.d.). housewarming. In *Cambridge Dictionary*. Retrieved September 10, 2023, from https://dictionary.cambridge.org/dictionary/english/housewarming

Cambridge Dictionary. (n.d.). whitewater rafting. In *Cambridge Dictionary*. Retrieved August 1, 2023, from https://dictionary.cambridge.org/dictionary/english/whitewater-rafting

Cambridge Dictionary. (n.d.). windsurfing. In *Cambridge Dictionary*. Retrieved August 1, 2023, from https://dictionary.cambridge.org/dictionary/english/windsurfing

Croft, W., & Cruse, D. A. (2004). *Cognitive linguistics*. Cambridge University Press.

Durán-Muñoz, I. (2016). Producing frame-based definitions. A case study. *Terminology*, 22(2), 223–249. https://doi.org/10.1075/term.22.2.04mun

Durán-Muñoz, I., & Jiménez-Navarro, E. L. (2021). Colocaciones verbales en el turismo de aventura: Estudio contrastivo inglés-español. In G. Corpas Pastor, M.ª R. Bautista Zambrana, & C. M. Hidalgo-Ternero (Eds.), *Sistemas fraseológicos en contraste: Enfoques computacionales y de corpus* (pp. 121–142). Comares.

Durán-Muñoz, I., & L'Homme, M.-C. (2020). Diving into English motion verbs from a lexico-semantic approach. A corpus-based analysis of adventure tourism. *Terminology*, 26(1), 33–59. https://doi.org/10.1075/term.00041.dur

Fillmore, C. J. (1976). Frame semantics and the nature of language. *Annals of the New York Academy of Sciences: Conference on the Origin and Development of Language and Speech*, *280*(1), 20–32. https://doi.org/10.1111/j.1749-6632.1976.tb25467.x

Fillmore, C. J. (1982). Frame semantics. In Linguistic Society of Korea (Ed.), *Linguistics in the morning calm* (pp. 111–137). Hanshin Publishing Company.

Fillmore, C. J., & Baker, C. (2010). A frames approach to semantic analysis. In B. Heine & H. Narrog (Eds.), *The Oxford handboook of linguistic analysis* (second edition) (pp. 313–340). Oxford University Press.

Hengeveld, K., & Mackenzie, L. (2008). *Functional discourse grammar. A typologically-based theory of language structure*. Oxford University Press.

Jiménez-Navarro, E. L. (This volume). Prepositional phrase collocations of motion verbs: A corpus-driven study in adventure tourism. In I. Durán-Muñoz & E. L. Jiménez-Navarro (Eds.), *Exploring the language of adventure tourism: A corpus-assisted approach*. Peter Lang.

Kilgarriff, A., Rychlý, P. S., Smrž, P., & Tugwell, D. (2004). The Sketch Engine. In G. Williams & S. Vessier (Eds.), *Proceedings of the 11th EURALEX International Congress* (pp. 105–116). Université de Bretagne-Sud, Faculté des lettres et des sciences humaines. https://doi.org/10.1007/s40607-014-0009-9

L'Homme, M.-C. (2012). Adding syntactico-semantic information to specialized dictionaries: An application of the FrameNet methodology. *Lexicographica*, *28*(1), 233–252. https://doi.org/10.1515/lexi.2012-0012

L'Homme, M.-C. (2018). Maintaining the balance between knowledge and the lexicon in terminology: A methodology based on Frame Semantics. *Lexicography*, *4*, 3–21.

Longman Dictionary of Contemporary English (n.d.). black-water rafting. In *Longman Dictionary of Contemporary English*. Retrieved August 1, 2023, from https://www.ldoceonline.com/dictionary/black-water-rafting

Lyons, J. (1977). *Semantics*. Volumes 1 and 2. Cambridge University Press.

Matthews, P. H. (1974). *Morphology: An introduction to the theory of word-structure*. Cambridge University Press.

Plag, I. (2003). *Word formation in English*. Cambridge University Press.

Plag, I. (2006). The variability of compound stress in English: Structural, semantic and analogical factors. *English Language and Linguistics*, *10*, 143–172.

Quirk, R., Greenbaum, S., Leech, G., & Svartvik, J. (1985). *A comprehensive grammar of the English language*. Longman.

Selkirk, E. O. (1982). *The syntax of words*. MIT Press.

Trask, R. L. (1993). *A dictionary of grammatical terms in linguistics*. Routledge.

Acknowledgements

This work has been carried out within the framework of the R&D project "DicoAdventure: diseño y desarrollo de un recurso electrónico especializado bilingüe (inglés, español) sobre el turismo de aventura a partir de marcos semánticos" (Ref. UCO-1380857-F), co-funded by the Operational Programme FEDER 2014-2020 and the Consejería de Economía, Conocimiento, Empresas y Universidad of the Andalusian regional government. The resource *DicoAdventure* (under construction) is available at http://olst.ling.umontreal.ca/dicoadventure/. Moreover, the research presented in this chapter was partially conducted within the framework of the project PGC2018-101214-B-Ioo "Researching conceptual metonymy in selected areas of grammar, discourse and sign language with the aid of the University of Córdoba Metonymy Database" (METGRADISL&BASE). I am indebted to Isabel Durán-Muñoz for embarking me in this challenging adventure and to the reviewers for their thorough revision and multiple suggestions. All remaining errors are my own.

Appendix: Relevance of the semantic role PATH in adventure tourism compound formation

Compound	Word 1 semantic class
bridge crossing	artefact_bridge
circuit trekking	artefact_circuit
oval track racing	artefact_oval track
wind tunnel indoor skydiving	artefact_place
pond swooping	artefact_pond
dirt road driving	artefact_road
road cycling	artefact_road
dirt track racing	artefact_tract
via ferrata climbing	artefact_via ferrata
via ferrata trekking	artefact_via ferrata
wall climbing	artefact_wall
canyon hiking	nature_canyon
canyon walking	nature_canyon
dirt biking	nature_dirt
dune bashing	nature_dune
fell running	nature_fell
ghyll scrambling	nature_ghyll
gorge paddling	nature_gorge
gorge scrambling	nature_gorge
gorge walking	nature_gorge
hill walking	nature_hill
mountain bicycling	nature_mountain
mountain biking	nature_mountain
mountain boarding	nature_mountain
mountain climbing	nature_mountain
mountain hiking	nature_mountain
(big) mountain skiing	nature_mountain
mountain trekking	nature_mountain
mountain walking	nature_mountain
Hampta Pass trekking	nature_name_corridor
Mount Rinjani trekking	nature_name_mount
Larapinta trail walking	nature_name_trail
Larke Pass trekking	nature_name_trail
Mustang trekking	nature_name_valley

Compound	Word 1 semantic class
Cerro Negro sand boarding	nature_name_volcano
Pacaya volcano boarding	nature_name_volcano
ocean kayaking	nature_ocean
ridge bashing	nature_ridge
river boarding	nature_river
river crossing	nature_river
river kayaking	nature_river
river rafting	nature_river
river running	nature_river
river surfing	nature_river
river tracing	nature_river
river trekking	nature_river
river tubing	nature_river
rock climbing	nature_rock
rock scrambling	nature_rock
sea kayaking	nature_sea
ice climbing	nature_substance_ice
ice skating	nature_substance_ice
sand boarding	nature_substance_sand
sand skiing	nature_substance_sand
snow boarding	nature_substance_snow
snow climbing	nature_substance_snow
snow skiing	nature_substance_snow
snow surfing	nature_substance_snow
snow trekking	nature_substance_snow
snow tubing	nature_substance_snow
wake boarding	nature_substance_wake
water kayaking	nature_substance_water
water rafting	nature_substance_water
water skiing	nature_substance_water
water surfing	nature_substance_water
whitewater kayaking	nature_substance_whitewater
whitewater rafting	nature_substance_whitewater
trail riding	nature_trail
trail running	nature_trail
tree climbing	nature_tree
treetop walking	nature_treetop

Compound	Word 1 semantic class
valley crossing	nature_valley
valley trekking	nature_valley
wake surfing	nature_wake
waterfall rappeling	nature_waterfall
waterfall swimming	nature_waterfall
bush bashing	plant_bush
vertical wind tunnel skydiving	property_artefact_tunnel
(vertical) wind tunnel skydiving	property_artefact_wind_tunnel
downhill skiing	property_direction
high diving	property_height
cross-country driving	property_place
cross-country skiing, x-country skiing	property_place
cross-country skydiving	property_place
downhill racing	property_place
off-road racing	property_place

Hallesche Sprach- und Textforschung
Language and Text Studies
Recherches linguistiques et textuelles

Herausgegeben von / Edited by / Dirigée par Alexander Brock & Daniela Pietrini

Bd./Vol. 1	Annette Schiller: Die présentatifs im heutigen Französisch. Eine funktionale Studie ihrer Vielfalt. 1992.
Bd./Vol. 2	Gertrud Bense (Hrsg.): Diachronie – Kontinuität – Impulse. Sprachwissenschaftliches Kolloquium Halle 1992. 1994.
Bd./Vol. 3	Wolfgang Boeck (Hrsg.): Sprache, Literatur und Landeskunde slavischer Völker. Funktionale Aspekte in der Beschreibung und Didaktik. 1994.
Bd./Vol. 4	Gertrud Bense (Hrsg.): Kommunikation und Grammatik. 1996.
Bd./Vol. 5	Gisela Hermann-Brennecke / Dietmar Schneider (Hrsg.): Dona Anglica. 120 Jahre Anglistik in Halle. 1997.
Bd./Vol. 6	Max Hans-Jürgen Mattusch: Vielsprachigkeit: Fluch oder Segen für die Menschheit? Zu Fragen einer europäischen und globalen Fremdsprachenpolitik. 1999.
Bd./Vol. 7	Christiane Schiller: Bilinguismus. Zur Darstellung eines soziolinguistischen Phänomens in der Literatur. Dargestellt an Beispielen der regionalen Literatur Preußisch-Litauens: Hermann Sudermann *Litauische Geschichten*, Ieva Simonaityte *Vilius Karalius*. 2000.
Bd./Vol. 8	Gertrud Bense: „Giedojam taw – Wir singen dir". Zur Textgeschichte der preußisch-litauischen Gesangbücher im 18. Jahrhundert. Mit besonderer Berücksichtigung der Liedersammlung von Fabian Ulrich Glaser (1688–1747) und ihrem Umfeld. 2001.
Bd./Vol. 9	Gertrud Bense / Gerhard Meiser / Edeltraud Werner (Hrsg.): August Friedrich Pott. Beiträge der Halleschen Tagung anlässlich des zweihundertsten Geburtstages von August Friedrich Pott (1802–1887). 2005.
Bd./Vol. 10	Julia Balakina: Anglicisms in Russian and German Blogs. A Comparative Analysis. 2011.
Bd./Vol. 11	Thomas Bremer/Annette Schiller (Hrsg.): Dialekt und Standardsprache in Italien und Europa. Edeltraud Werner zum 60. Geburtstag. 2012.
Bd./Vol. 12	Anne Ammermann / Alexander Brock / Jana Pflaeging / Peter Schildhauer (eds.): Facets of Linguistics. Proceedings of the 14th Norddeutsches Linguistisches Kolloquium 2013 in Halle an der Saale. 2013.
Bd./Vol. 13	Anja Neuber: Perspektiven des Friaulischen. Eine soziolinguistische Untersuchung am Beispiel junger Erwachsener. 2015.
Bd./Vol. 14	Peter Schildhauer: The Personal Weblog. A Linguistic History. 2016.
Bd./Vol. 15	Alexander Brock / Peter Schildhauer (eds.): Communication Forms and Communicative Practices. New Perspectives on Communication Forms, Affordances and What Users Make of Them. 2017.
Bd./Vol. 16	Alexander Brock / Jana Pflaeging / Peter Schildhauer (eds.): Genre Emergence. Developments in Print, TV and Digital Media. 2019.
Bd./Vol. 17	Björn Langkopf: Autonomes E-Learning. Effizienz – Didaktik – Perspektiven. 2019.
Bd./Vol. 18	John Marcus Sommer: English and French Online Comments. A Text Linguistic Comparison of Popular Science Magazines. 2020.
Bd./Vol. 19	José Luis Oncins Martínez (ed.): Current Trends in Corpus Linguistics. 2020.

Bd./Vol. 20 Ramón Martí Solano / Pablo Ruano San Segundo (eds.): Anglicisms and Corpus Linguistics. Corpus-Aided Research into the Influence of English on European Languages. 2021.

Bd./Vol. 21 Cristina Fernández-Alcaina: The Competition of Word-Formation Processes in the Derivational Paradigm of Verbs. Diasynchronic Evidence for the Profile and Resolution of Competition in English. 2021.

Bd./Vol. 22 Alexander Brock / Janet Russell / Peter Schildhauer / Merle Willenberg (eds.): Participation & Identity. Empirical Investigations of States and Dynamics. 2022.

Bd./Vol. 23 Daniela Pietrini (a cura di): Lingua e discriminazione. Studi diacronici, lessicali e discorsivi. 2023.

Bd./Vol. 24 Isabel Durán-Muñoz / Eva Lucía Jiménez-Navarro (eds.): Exploring the Language of Adventure Tourism. A Corpus-Assisted Approach. 2024.

www.peterlang.com

www.ingramcontent.com/pod-product-compliance
Ingram Content Group UK Ltd.
Pitfield, Milton Keynes, MK11 3LW, UK
UKHW041923210426
5322IPUK00002B/34